AF856339

HVAC and Refrigeration Six-Minute Problems

Second Edition

Keith E. Elder, PE

PPI®
PPI2PASS.COM

Professional Publications, Inc. • Belmont, California

Benefit by Registering This Book with PPI

- Get book updates and corrections.
- Hear the latest exam news.
- Obtain exclusive exam tips and strategies.
- Receive special discounts.

Register your book at **ppi2pass.com/register**.

Report Errors and View Corrections for This Book

PPI is grateful to every reader who notifies us of a possible error. Your feedback allows us to improve the quality and accuracy of our products. You can report errata and view corrections at **ppi2pass.com/errata**.

HVAC AND REFRIGERATION SIX-MINUTE PROBLEMS
Second Edition

Current release of this edition: 11

Release History

date	edition number	revision number	update
Aug 2015	2	9	Minor cover updates.
Jul 2016	2	10	Minor corrections. Minor formatting and pagination changes.
Aug 2017	2	11	Minor corrections. Minor cover updates.

Printed in the United States of America.

PPI
1250 Fifth Avenue, Belmont, CA 94002
(650) 593-9119
ppi2pass.com

ISBN: 978-1-59126-552-8

Library of Congress Control Number: 2008936922

F E D C B A

Table of Contents

About the Author

Keith Elder, PE, is principal advisor for the Seattle office of Coffman Engineers. He also serves on the faculty at the University of Washington, where he teaches HVAC design and PE exam preparation for the mechanical engineering department and the Engineering Professional Programs department. Mr. Elder's career has included the design and analysis of commercial HVAC systems, the investigation and correction of system performance problems, and participation in the development of local and state energy codes.

A passion to learn about energy and its efficient use, sparked by the energy crisis of the early 1970s, led Mr. Elder into engineering school and eventually to studies in mechanical engineering at the University of Washington. After related internships in aerospace and manufacturing, he returned to complete his bachelor's degree in mechanical engineering in 1981. After graduation, Mr. Elder began a career as a consultant in HVAC engineering. In these early years, he focused on design principles and practice, but his first passion remained efficiency in energy processes, and increasingly this led to work in HVAC systems analysis and computer modeling of building performance.

Mr. Elder became a registered mechanical engineer in the state of Washington in 1986. By this time he was an experienced lecturer, and this led to an opportunity to teach HVAC design for the Engineering Professional Programs department at the University of Washington. He still teaches fluid dynamics, hydraulic machines, and HVAC twice a year as part of a refresher course given by the department for those studying for the PE exam.

Mr. Elder soon began teaching HVAC design in the undergraduate program as well, and in 1988 was appointed to the faculty of the mechanical engineering department, where he continues to teach.

In 1990, Mr. Elder and his colleague Don Iverson started a mechanical engineering consulting firm called Iverson Elder Inc. After nearly ten years of successful practice, they merged with Coffman Engineers, becoming principals in that firm. Mr. Elder continued to work in energy and systems analysis, and also enjoyed the opportunity his new position brought him of mentoring and training engineers within the company. He became managing partner of Coffman's Seattle office in 2003.

In recent years, Mr. Elder's strong interest in the professional development of engineers has moved him to become more active in the area of engineering ethics. He has worked with the Gonzaga Institute of Ethics, and has published a paper on ethics education in the journal *Science and Engineering Ethics*.

Mr. Elder and his wife of 25 years, Patricia, are longtime residents of Seattle. They are active in their church and enjoy international travel in their free time.

Preface and Acknowledgments

The Principles and Practice of Engineering examination (PE exam) for mechanical engineering is prepared by the National Council of Examiners for Engineering and Surveying (NCEES). The exam is developed from sample problems submitted by educators and professional engineers representing consulting, government, and industry. PE exams are designed to test examinees' understanding of both theoretical and practical engineering concepts. Problems from past exams are not available from NCEES or any other sources. However, NCEES does identify the general subject areas covered in the exam.

The topics covered in *HVAC and Refrigeration Six-Minute Problems* are the same as those identified by NCEES for the HVAC depth module of the mechanical PE exam. These topics include psychrometrics, cooling and heating cycles, humidification and dehumidification, energy balances, boilers and furnaces, evaporators and chillers, energy recovery, vibration control, acoustics, the codes and standards of the American Society of Heating, Refrigerating, and Air-Conditioning Engineers (ASHRAE), and many others.

One of the best preparations for the exam is simply to get back into the habit of formal problem solving. Keeping up the drill of solving problems is in some respects just as important as finding the right problems to solve.

This book will help you with both. The problems are representative of the types and breadth of those you will face on the PE exam. Both conceptual and practical problems at varying levels of difficulty are included. The problems you'll encounter on the exam probably won't be exactly like the ones presented here. But reviewing these problems and solutions will increase your familiarity with the form, content, and solution methods of the exam problems. This preparation will help you considerably during the exam.

The problems and solutions have been carefully preparedand reviewed to ensure that they are appropriate and understandable, and that they are solved correctly. If you find errors or discover an alternative, more efficient way to solve a problem, please bring it to PPI's attention so that your suggestions can be incorporated into future editions. You can report errors and keep up with changes made to this book, as well as changes to the exam, by going to PPI's website at **ppi2pass.com/errata**.

For the second edition, the problems have been reorganized. The order of topics in this book now matches the organization of subject areas as given in the current NCEES specifications for the mechanical PE exam, which became effective in October 2008. We hope this will help you focus your study to the needs of the current exam.

I'd like to thank the many persons in the editorial and production departments at PPI who contributed to the successful publication of this book. They have been great to work with, and are dedicated to making these publications the best they can possibly be. Patient and professional are the two words they bring to mind. I am especially grateful to Aline McGee, for her gentle recruiting and patient introduction to PPI, and to Scott Marley, who has mastered the balance between maintaining consistent standards and allowing for authorial idiosyncrasy. Thanks are also due to Amy Schwertman, who drew the illustrations, Kate Hayes, who typeset and laid out the text, Matt Gordon, for his thorough technical review, Matthew Breault, who carefully checked the calculations, editorial manager Sarah Hubbard, and production manager Cathy Schrott. They contributed many suggestions and improvements to the content and presentation of this book.

I'd especially like to acknowledge my professional colleagues at Coffman Engineers who continue to stimulate me with their intellectual vitality and passion for their work. Thanks to Professor Creighton Depew, who helped me establish the academic foundation for my professional career and has honored me with his friendship in the years since. And thanks to my father, David, who saw before I did that I was destined to be an engineer, and encouraged me in that pursuit. Finally, words can't express my gratitude to my wife, Pat, the love of my life. In both sickness and health, you have been the instigator and inspiration behind every worthwhile activity.

Keith E. Elder, PE

Introduction

EXAM FORMAT

The Principles and Practice of Engineering examination (PE exam) in mechanical engineering is an 8-hour exam divided into a morning and an afternoon session. The morning session is known as the "breadth" exam and the afternoon is known as the "depth" exam.

The morning session includes 40 problems from all three areas of mechanical engineering: HVAC and refrigeration, mechanical systems and materials, and thermal and fluids systems. The three areas are roughly equally represented. As the "breadth" designation implies, morning session problems are general in nature and wide-ranging in scope.

The afternoon session allows the examinee to select a "depth" exam module from one of the three subdisciplines. The 40 problems included in the afternoon session require more specialized knowledge than those in the morning session.

All problems from both the morning and afternoon sessions are multiple choice. They include a problem statement with all required defining information, followed by four logical choices. Only one of the four options is correct. Problems are generally self-contained and independent, so an incorrect choice on one problem typically will not carry over to subsequent problems.

Topics and the approximate distribution of problems on the afternoon session of the mechanical HVAC and refrigeration exam are as follows.

PRINCIPLES: ABOUT 55% OF EXAM PROBLEMS

- Thermodynamics: cycles, properties, compression processes
- Psychrometrics: heating and cooling cycles, humidification and dehumidification, heating and cooling loads
- Heat transfer
- Fluid mechanics
- Compressible flow
- Energy balances

APPLICATIONS: ABOUT 45% OF EXAM PROBLEMS

- Equipment and components: cooling towers and fluid coolers, boilers and furnaces, condensers, pumps, compressors, fans, evaporators and chillers, cooling and heating coils, control systems components, refrigerants, refrigeration components
- Systems: air distribution, fluid distribution, refrigeration, energy recovery
- Supportive knowledges: codes and standards, air quality and ventilation, vibration control, acoustics, economic analysis, electrical concepts

For further information and tips on how to prepare for the mechanical engineering PE exam, consult the *Mechanical Engineering Reference Manual* or PPI's website, at **ppi2pass.com/mefaq**.

THIS BOOK'S ORGANIZATION

HVAC and Refrigeration Six-Minute Problems is organized into two sections. The first section, Breadth Problems, presents 20 problems in HVAC and refrigeration of the type that would be expected in the morning part of the mechanical engineering PE exam. The second section, Depth Problems, presents 65 problems typical of the afternoon part of this exam. Each of these two sections of the book is further subdivided into principles and applications.

Most of the problems are quantitative, requiring calculations to arrive at a correct solution. A few are nonquantitative. Some problems will require a little more than 6 minutes to answer and others a little less. On average, you should expect to complete 80 problems in 480 minutes (8 hours), or spend 6 minutes per problem.

HVAC and Refrigeration Six-Minute Problems does not include problems related directly to mechanical systems and materials, nor to thermal and fluids systems, although problems from these subdisciplines will be included in the HVAC and refrigeration exam, particularly in the morning session. Other books in the *Six-Minute* series provide problems for review in these areas.

HOW TO USE THIS BOOK

Each problem statement in this book, with its supporting information and answer choices, is presented in the same format as the problems encountered on the PE exam. The solutions explain step by step how the answer is logically derived, to help you follow the reasoning and to provide examples of how you may want to approach your solutions as you take the PE exam.

Each problem includes a hint to provide direction in solving the problem. In addition to the correct solution, you will find an explanation of the faulty reasoning leading to the three incorrect answer choices. The incorrect answers are chosen to show some common mistakes made when solving each type of problem. These may be simple mathematical errors, such as failing to square a term in an equation, or more serious errors, such as using the wrong equation.

To optimize your study time and obtain the maximum benefit from the practice problems, consider the following suggestions.

1. Complete an overall review of the problems and identify the subjects that you are least familiar with. Work a few of these problems to assess your general understanding of the subjects and to identify your strengths and weaknesses.

2. Locate and organize relevant resource materials. (See the references section of this book as a starting point.) As you work problems, some of these resources will emerge as more useful to you than others. These are what you will want to have on hand when taking the PE exam.

3. Work the problems in one subject area at a time, starting with the subject areas that you have the most difficulty with.

4. When possible, work problems without using the hint. Always attempt your own solution before looking at the solutions provided in the book. Use the solutions to check your work or to provide guidance in finding solutions to the more difficult problems. Use the incorrect solutions to help identify pitfalls and to develop strategies to avoid them.

5. Use each subject area's solutions as a guide to understanding general problem-solving approaches. Although problems identical to those presented in this book will not be encountered on the PE exam, the approach to solving problems will be the same.

Solutions presented for each example problem may represent only one of several methods for obtaining a correct answer. Although we have tried to prepare problems with unique solutions, alternative problem-solving methods may occasionally produce a different, but nonetheless appropriate, answer.

References

The minimum recommended library for the mechanical PE exam is PPI's *Mechanical Engineering Reference Manual.* You may also find the following references helpful in completing some of the problems in *HVAC and Refrigeration Six-Minute Problems.*

American National Standards Institute. *ANSI Z21.10/ CSA 4: Gas Water Heaters.*

American National Standards Institute. *ANSI Z21.47/ CSA 2.3: Gas-Fired Central Furnaces.*

American Society of Heating, Refrigerating, and Air-Conditioning Engineers. *ASHRAE Handbook: Fundamentals.*

American Society of Heating, Refrigerating, and Air-Conditioning Engineers. *ASHRAE Handbook: Refrigeration.*

American Society of Heating, Refrigerating, and Air-Conditioning Engineers. *ASHRAE Handbook: HVAC Applications.*

American Society of Heating, Refrigerating, and Air-Conditioning Engineers. *ASHRAE Handbook: HVAC Systems and Equipment.*

American Society of Heating, Refrigerating, and Air-Conditioning Engineers. *ANSI/ASHRAE Standard 15: Safety Standard for Refrigerating Systems.*

American Society of Heating, Refrigerating, and Air-Conditioning Engineers. *ANSI/ASHRAE Standard 34: Designation and Safety Classification of Refrigerants.*

American Society of Heating, Refrigerating, and Air-Conditioning Engineers. *ASHRAE Standard 55: Thermal Environmental Conditions for Human Occupancy.*

American Society of Heating, Refrigerating, and Air-Conditioning Engineers. *ANSI/ASHRAE Standard 62.1: Ventilation for Acceptable Indoor Air Quality.*

American Society of Heating, Refrigerating, and Air-Conditioning Engineers. *ANSI/ASHRAE Standard 62.2: Ventilation and Acceptable Indoor Air Quality in Low-Rise Residential Buildings.*

American Society of Heating, Refrigerating, and Air-Conditioning Engineers. *ASHRAE Standard 84: Method of Testing Air-to-Air Heat Exchangers.*

American Society of Heating, Refrigerating, and Air-Conditioning Engineers. *ANSI/ASHRAE/IESNA Standard 90.1: Energy Standard for Buildings Except Low-Rise Residential Buildings, I-P Edition.*

Cooling Technology Institute. *CTI Standard 201: Standard for the Certification of Water Cooling Tower Thermal Performance.*

Crane Valve Group. *Crane Technical Paper No. 410: Flow of Fluids Through Valves, Fittings, and Pipes.*

Hydraulic Institute. *Engineering Data Book.*

International Association of Plumbing and Mechanical Officials. *Uniform Mechanical Code.*

Nomenclature

A	annual cost	\$/yr	\$/yr
A	area	ft^2	m^2
ADPI	air diffusion performance index	–	–
AFUE	annual fuel utilization efficiency	–	–
BF	bypass factor	–	–
BHP	boiler or brake horsepower	hp	W
c	specific heat	Btu/lbm-°F	J/kg·°C
C	cost	\$	\$
C	fitting coefficient	–	–
C	thermal capacity	Btu/hr	W
COP	coefficient of performance	–	–
CRF	condensation resistance factor	–	–
d	diameter	ft	m
E	energy usage	kW-hr/yr	kW·h/yr
EER	energy efficiency rating	–	–
f	frequency	Hz	Hz
f	friction factor	–	–
F	factor	–	–
g	gravitational acceleration	ft/sec^2	m/s^2
g_c	gravitational constant	$ft\text{-}lbm/lbf\text{-}sec^2$	n.a.
h	enthalpy	Btu/lbm	J/kg
h	head, head loss,or height	ft	m
h	heat transfer coefficient	$Btu/hr\text{-}ft^2\text{-}°F$	$W/m^2·K$
H	humidification load	lbm/sec	kg/s
I	current	A	A
L	length	ft	m
L	sound level	dB	dB
m	mass	lbm	kg
$\dot{m}$	mass flow rate	lbm/sec	kg/s
n	number of units	–	–
n	rotational speed	rpm	rpm
N	number or fractional part	–	–
NPSHA	net positive suction head available	ft	m
$oa_{\%}$	percentage of outside air	%	%
p	pressure	lbf/ft^2	Pa
P	power	hp	W
PF	power factor	–	–
q	heat	Btu	J
q	heat per unit mass	Btu/lbm	J/kg
Q	flow rate	gal/sec	L/s
Q	heat transfer rate	Btu/hr	W
r	energy cost	\$/kW-hr	\$/kW·h
r	distance to observation point	ft	m
R	thermal resistance	$hr\text{-}ft^2\text{-}°F/Btu$	$m^2·°C/W$
s	entropy	Btu/lbm-°F	J/kg·°C
SC	shading coefficient	–	–
SCL	solar cooling load factor	$Btu/hr\text{-}ft^2$	W/m^2
SHR	sensible heat ratio	–	–
SP	static pressure	in wg	Pa
t	time	sec	s
T	temperature	°F	°C
T	throw	ft	m
T	transmissibility	–	–
TP	total pressure	in wg	Pa
U	overall coefficient of heat transfer	$Btu/hr\text{-}ft^2\text{-}°F$	$W/m^2·°C$
v	velocity	ft/sec	m/s
V	voltage	V	V
V	volume	ft^3	m^3
$\dot{V}$	volumetric flow rate	ft^3/sec	m^3/s
VP	velocity pressure	in wg	Pa
w	width	ft	m
w_p	evaporation rate	lbm/sec	kg/s
$w_{\%}$	percentage of water content	%	%
W	work	Btu/lbm	J/kg
x	quality	–	–
Y	latent heat to change water to vapor	Btu/lbm	J/kg
z	elevation	ft	m

Symbols

δ	deflection	ft	m
ϵ	effectiveness	–	–
η	efficiency	–	–
ρ	density	lbm/ft^3	kg/m^3
τ	time constant	sec	s
υ	specific volume	ft^3/lbm	m^3/kg
ϕ	relative humidity	–	–
ω	humidity ratio	lbm/lbm	kg/kg

Subscripts

0	initial
a	air, dry air
abs	absolute
atm	atmospheric
aux	auxiliary
c	coil, condensation
cap	capacity
comp	compressed
d	disturbing
db	dry-bulb
dp	dew point
eq	equivalent
f	fitting, fluid (liquid), friction, fusion
fg	vaporization (fluid to gas)
g	gas (vapor)
hyd	hydrostatic
i	inside
m	middle
M	added by hydraulic machine
n	natural
o	operative, other, outside
oa	outside air
p	constant pressure, pool, pressure
r	mean radiant, radiation
s	sensible
sa	supply air
sat	saturation
st	static
t	transmitted
vp	vapor pressure
w	water vapor, window
W	power
z	potential, static

1 Breadth

PRINCIPLES

PROBLEM 1

Refrigerant-22 is isentropically compressed from a saturated vapor condition at 68.5 psig to 185 psig. The total change in temperature of the vapor is most nearly

(A) 60°F

(B) 70°F

(C) 80°F

(D) 90°F

Hint: Using a pressure-enthalpy chart for R-22, locate the entropy of the saturated vapor curve. Then follow a constant-entropy process line until it intercepts the final pressure condition line.

PROBLEM 2

A room is conditioned to 75°F and 50% relative humidity. A fan system combines 5000 ft^3/min of recirculated air from the room with 2500 ft^3/min of outdoor air at 90°F dry-bulb and 75°F wet-bulb temperature. The mixed air enters through a chilled-water cooling coil as shown.

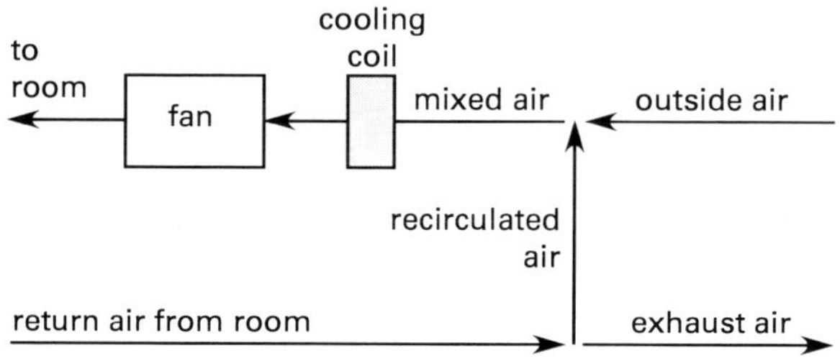

The enthalpy of the air entering the cooling coil is

(A) 28 Btu/lbm

(B) 32 Btu/lbm

(C) 35 Btu/lbm

(D) 39 Btu/lbm

Hint: The psychrometric properties of the mixed air are a weighted interpolation of the outside air and recirculated air volume flows.

PROBLEM 3

An HVAC system conditions a library containing rare books. The system is designed to maintain the space at precisely 75°F dry-bulb and 61°F wet-bulb temperature. The vapor pressure of the space, in inches of mercury, is most nearly

(A) 0.0080 in Hg

(B) 0.20 in Hg

(C) 0.40 in Hg

(D) 13 in Hg

Hint: Vapor pressure is the term used for the partial pressure of the water vapor in air. By definition, relative humidity is the partial pressure of the water vapor divided by the saturation pressure.

PROBLEM 4

Which of these psychrometric processes can successfully condition air from 81°F dry-bulb temperature and 45% relative humidity to 60°F dry-bulb temperature and 51°F dew point?

(A) sensible cooling and latent heating

(B) sensible cooling and latent cooling

(C) sensible heating and latent heating

(D) latent cooling and sensible heating

Hint: Sensible processes change dry-bulb temperature. Latent processes change humidity.

PROBLEM 5

As part of the heat loss calculation for a storage facility, it is necessary to determine the U-factor for the wall shown.

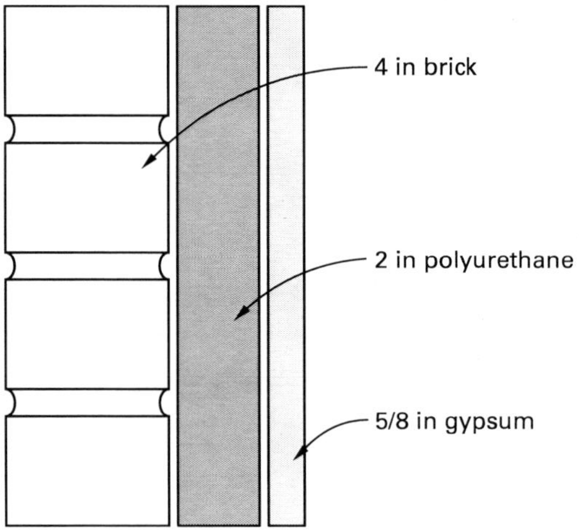

The constituent layers have the following thermal resistivities (in units of hr-ft^2-°F/Btu-in).

face brick	0.10
polyurethane insulation	6.0
gypsum board	0.90

The air films have the following resistances (in units of hr-ft^2-°F/Btu).

indoor air film	0.68
outdoor air film	0.17

The overall coefficient of heat transfer, U, for the wall is most nearly

(A) 0.072 Btu/hr-ft^2-°F

(B) 0.077 Btu/hr-ft^2-°F

(C) 0.13 Btu/hr-ft^2-°F

(D) 12 Btu/hr-ft^2-°F

Hint: The U-factor is the reciprocal of the total thermal resistance of the wall.

PROBLEM 6

A compressor's piston travels to 20% of its original volume. It is used to compress 70°F air from atmospheric pressure to 100 psig. Assuming that the air behaves as an ideal gas, the temperature of the compressed air is most nearly

(A) 110°F

(B) 260°F

(C) 340°F

(D) 370°F

Hint: For an ideal gas, the relationship between pressure, volume, and temperature is $p_1V_1/T_1 = p_2V_2/T_2$.

PROBLEM 7

A chiller uses chilled water as a distribution medium. The refrigeration cycle for this chiller operates at an evaporating temperature of 40°F and a condensing temperature of 120°F. The refrigerant is superheated prior to compression and subcooled before leaving the condenser. Which of the following parameters best delimits the net system refrigerating effect?

(A) from subcooled liquid enthalpy to the enthalpy at the superheated condition

(B) from the enthalpy at saturated liquid to the enthalpy at the superheated condition

(C) from subcooled liquid enthalpy to the saturated vapor enthalpy

(D) from the enthalpy at saturated liquid to the saturated vapor enthalpy

Hint: The net refrigeration effect is defined by ASHRAE as the rate at which heat is removed by the primary refrigerant from the cooling medium that transmits the refrigerating effect.

PROBLEM 8

A single-stage refrigeration cycle operates with refrigerant-123. Assuming an isentropic process, no subcooling, 40°F superheat, a condensing temperature of 260°F, and an evaporating temperature of 100°F, the coefficient of performance is most nearly

(A) 1.6

(B) 1.7

(C) 2.1

(D) 4.4

Hint: The coefficient of performance of a refrigeration cycle is the net refrigeration effect divided by the compressor work.

PROBLEM 9

800 lbm of water is heated to a state where the pressure is 20 psig and only 100 lbm remains in its liquid state. The entropy of the final mixture is most nearly

(A) 0.50 Btu/lbm-°R

(B) 1.5 Btu/lbm-°R

(C) 1.6 Btu/lbm-°R

(D) 1.7 Btu/lbm-°R

Hint: Properties of two-phase mixtures can be determined by using the quality of the mixture to find actual properties between the state points at saturated conditions.

APPLICATIONS

PROBLEM 10

Which of the following cycles would be appropriate for providing cooling to a commercial building?

(A) a Rankine cycle

(B) a Brayton cycle

(C) a vapor compression cycle

(D) an Otto cycle

Hint: Consider which of the cycles listed involves refrigeration.

PROBLEM 11

During the winter, the indoor environment in a hospital is maintained at 74°F dry-bulb temperature and 40% relative humidity. New windows are to be designed to avoid condensation. On a "design day," the local outdoor temperature is −12°F. The minimum condensation resistance factor (CRF) that should be specified for the windows is most nearly

(A) 0.42

(B) 0.58

(C) 0.70

(D) 0.80

Hint: The condensation resistance factor is the ratio of the dew point/outdoor temperature difference to the total indoor/outdoor temperature difference.

PROBLEM 12

A local ordinance requires that the 140°F condensate resulting from utility-provided steam (with no condensate return to the utility) be cooled before it is discharged to the sewer. A local inventor is developing a machine that will not only do this, but will also generate power by making use of the temperature difference between the condensate and the 50°F average sewer system temperature. The maximum theoretical efficiency the inventor can expect is most nearly

(A) 15%

(B) 36%

(C) 65%

(D) 85%

Hint: The quickest way to estimate a maximum theoretical benefit is to calculate the Carnot efficiency for a power cycle operating between two constant temperature reservoirs.

PROBLEM 13

The chapel in a hospital is heated with a steam radiator. The heating load is 47,000 Btu/hr. The minimum rate at which 5 psig saturated steam must be delivered to the radiator is most nearly

(A) 41 lbm/hr

(B) 47 lbm/hr

(C) 48 lbm/hr

(D) 49 lbm/hr

Hint: Assume that the radiator output is liquid, so that the capacity of the steam for heating is based on the mass flow rate and the latent energy of the steam.

PROBLEM 14

A chiller is rated to have a capacity of 600 tons and has a coefficient of performance of 5.5 at standard ARI conditions. Electricity costs $0.09/kW-hr. The cost of one hour of chiller operation is most nearly

(A) $10

(B) $35

(C) $190

(D) $350

Hint: The designation "ton" is an industry parameter equal to 12,000 Btu/hr.

PROBLEM 15

Dehumidification of moist air may be accomplished by three of the following methods. Which is the exception?

(A) solid sorption

(B) use of liquid desiccant

(C) air compression

(D) aeration

Hint: Air loses its ability to hold moisture at higher pressures.

PROBLEM 16

An air distribution system is being designed for a buffet-style restaurant. The solar load on the walls and windows is 50,000 Btu/hr and internal lighting loads are 20 kW. People contribute 30,000 Btu/hr of sensible heat and 40,000 Btu/hr of latent, and the buffet steam tables contribute 105,000 Btu/hr. The air handling equipment should be selected based upon a sensible load of most nearly

(A) 100,000 Btu/hr

(B) 150,000 Btu/hr

(C) 190,000 Btu/hr

(D) 250,000 Btu/hr

Hint: Consider the sensible and latent components, if any, of the steam table load.

PROBLEM 17

A cooling coil has a bypass factor of 0.20 and operates with an apparatus dew-point temperature of 40°F. Air is introduced into the coil at conditions of 80°F dry-bulb and 65°F wet-bulb temperature. The psychrometric state point of the air leaving the coil is most nearly

(A) 40°F dry-bulb, 40°F wet-bulb

(B) 48°F dry-bulb, 46°F wet-bulb

(C) 57°F dry-bulb, 57°F wet-bulb

(D) 72°F dry-bulb, 61°F wet-bulb

Hint: The answer lies along the straight line that connects the state point of the air entering the coil with the apparatus dew-point (coil) temperature.

PROBLEM 18

One kind of refrigeration process operates by evaporating refrigerant and then combining it into another medium to be expelled later by a heating process. This is commonly referred to as

(A) adsorption dehumidification

(B) cascade refrigeration

(C) direct expansion refrigeration

(D) absorption refrigeration

Hint: Dehumidification may be an application of a refrigeration process, but it is not itself a refrigeration process.

PROBLEM 19

A building 350 ft tall is heated to 70°F when the outside temperature is 10°F. Assume the neutral plane is at the building's midheight. The pressure difference across the entrance doors at the base of the building, in inches of water gage, is most nearly

(A) −0.34 in wg

(B) −1.8 in wg

(C) −11 in wg

(D) −50 in wg

Hint: The differential pressure across the doorway will be proportional to the difference in density of two air columns, one with its density calculated at 10°F and the other with its density calculated at 70°F.

PROBLEM 20

The air in the bladder-type expansion tank shown has been pressurized to 15 psig.

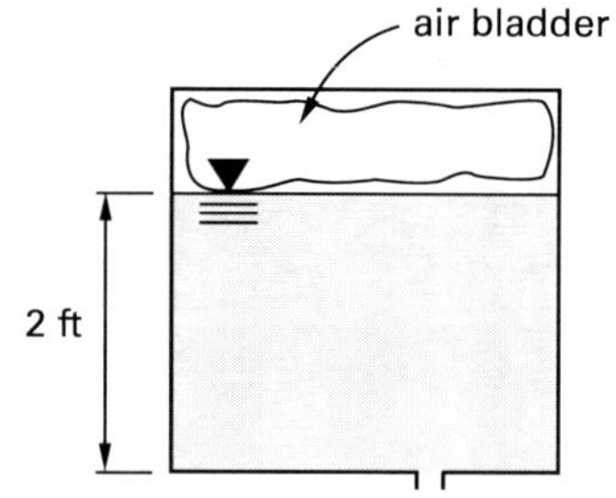

The total pressure at the bottom exit of the tank is most nearly

(A) 16 psia

(B) 31 psia

(C) 60 psia

(D) 150 psia

Hint: The solution requires adding the absolute pressure due to the air bladder to the hydrostatic pressure of the 2 ft of water.

SOLUTION 1

From the pressure-enthalpy chart for R-22, the absolute pressure of the saturated refrigerant vapor is

$$\begin{aligned} p_{\text{abs}} &= p_{\text{gage}} + p_{\text{atm}} \\ &= 68.5\ \frac{\text{lbf}}{\text{in}^2} + 14.7\ \frac{\text{lbf}}{\text{in}^2} \\ &= 83.2\ \text{psia} \end{aligned}$$

Similarly, the absolute pressure of the compressed refrigerant vapor is

$$\begin{aligned} p_{\text{abs}} &= 185\ \frac{\text{lbf}}{\text{in}^2} + 14.7\ \frac{\text{lbf}}{\text{in}^2} \\ &= 199.7\ \text{psia} \end{aligned}$$

From the table of R-22 properties, saturated vapor at approximately 83 psia has an entropy of approximately 0.22 Btu/lbm-°R and a temperature of approximately 40°F. Locate that point on the R-22 pressure-enthalpy diagram, and follow the process along a line of constant entropy until it intercepts 200 psia. The temperature at this point is approximately 120°F. The change in temperature is

$$\begin{aligned} \Delta T &= T_{\text{comp}} - T_{\text{sat}} \\ &= 120°\text{F} - 40°\text{F} \\ &= 80°\text{F} \end{aligned}$$

The answer is (C).

Why Other Options Are Wrong

(A) This incorrect answer results from following the saturated vapor line (rather than the constant entropy line) to its interception with 200 psia.

(B) This answer is wrong because the conversion from gage to absolute pressure was omitted for the compressed vapor.

(D) This answer is wrong because the conversion from gage to absolute pressure units was omitted for the saturated vapor.

SOLUTION 2

The mixed-air enthalpy can be determined graphically on the psychrometric chart, or by the ratio of the quantities of outside and recirculated air. From the psychrometric chart, the enthalpy for outside air at 90°F dry-bulb and 75°F wet-bulb is 38.5 Btu/lbm of dry air. The enthalpy for air returning from the room at 75°F dry-bulb and 50% relative humidity is 28.1 Btu/lbm of dry air.

$$\begin{aligned} h_{\text{mixed}} &= \left(\frac{\dot{V}_{\text{recirc}}}{\dot{V}_{\text{recirc}} + \dot{V}_{\text{outside}}}\right) h_{\text{recirc}} \\ &\quad + \left(\frac{\dot{V}_{\text{outside}}}{\dot{V}_{\text{recirc}} + \dot{V}_{\text{outside}}}\right) h_{\text{outside}} \\ &= \left(\frac{5000\ \frac{\text{ft}^3}{\text{min}}}{5000\ \frac{\text{ft}^3}{\text{min}} + 2500\ \frac{\text{ft}^3}{\text{min}}}\right)\left(28.1\ \frac{\text{Btu}}{\text{lbm}}\right) \\ &\quad + \left(\frac{2500\ \frac{\text{ft}^3}{\text{min}}}{5000\ \frac{\text{ft}^3}{\text{min}} + 2500\ \frac{\text{ft}^3}{\text{min}}}\right)\left(38.5\ \frac{\text{Btu}}{\text{lbm}}\right) \\ &= 31.6\ \text{Btu/lbm} \quad (32\ \text{Btu/lbm}) \end{aligned}$$

The answer is (B).

Why Other Options Are Wrong

(A) This incorrect answer is based on the state point of the return air conditions of 75°F dry-bulb and 50% relative humidity and does not reflect the mixing that must take place.

(C) This incorrect answer results from applying the level rule backward, reversing the ratio of recirculated to outside air (a common mistake).

(D) This incorrect answer is based on the state point of the outdoor conditions, 90°F dry-bulb and 75°F wet-bulb and does not reflect the mixing that must take place.

SOLUTION 3

Relative humidity is the partial pressure of the water vapor divided by the saturation pressure.

$$\begin{aligned} \phi &= \frac{p_w}{p_{\text{sat}}} \\ p_w &= \phi p_{\text{sat}} \end{aligned}$$

From the psychrometric chart, the relative humidity of air at the conditions given is approximately 45%.

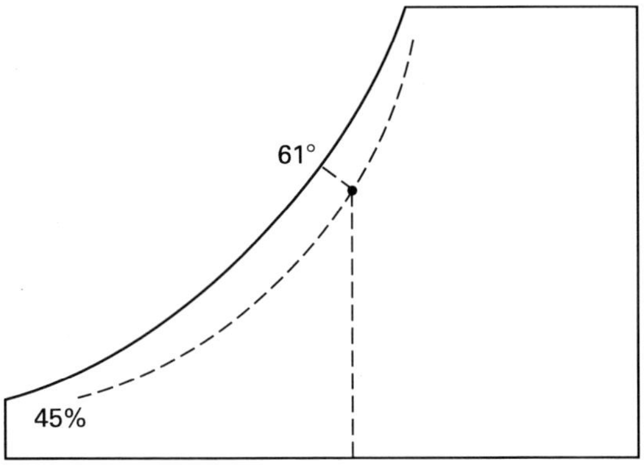

The saturation pressure of the water vapor corresponding to the air temperature can be obtained from the steam tables. At 75°F, the saturation pressure is

$$p_{sat} = 0.4299 \text{ psi}$$

The vapor pressure can be expressed as the partial pressure of the water vapor

$$\begin{aligned} p_w &= \phi p_{sat} \\ &= (0.45)\left(0.4299 \ \frac{\text{lbf}}{\text{in}^2}\right)\left(\frac{2.036 \text{ in Hg}}{1 \ \frac{\text{lbf}}{\text{in}^2}}\right) \\ &= 0.3939 \text{ in Hg} \quad (0.40 \text{ in Hg}) \end{aligned}$$

The answer is (C).

Why Other Options Are Wrong

(A) This answer is wrong because the humidity ratio, rather than partial pressure, was taken from the psychrometric chart. This could work if converted properly, but requires the use of the Carrier equation.

(B) This answer is wrong because units were not converted from pounds per square inch to inches of mercury.

(D) This answer is wrong because relative humidity was taken as the fraction of total air pressure in inches of mercury, rather than the fraction of saturation pressure at the conditions given.

SOLUTION 4

The initial and final states of the air are plotted on the psychrometric chart shown. The process from 81°F dry-bulb to 60°F dry-bulb temperature is a sensible cooling process. The process from 45% relative humidity (at 81°F dry-bulb) to 51°F dew point (at 60°F dry-bulb) is a latent cooling process.

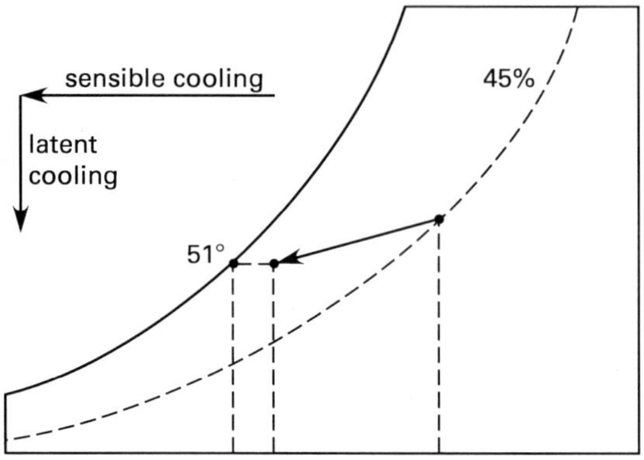

The answer is (B).

Why Other Options Are Wrong

(A) This answer correctly identifies a sensible cooling process but incorrectly states a latent *heating* process, which would require an *increase* in humidity, moving upward rather than downward on the chart.

(C) This answer incorrectly identifies a sensible heating process, which would require an *increase* in temperature, moving to the right on the chart rather than to the left.

(D) This answer correctly identifies a latent cooling process but incorrectly states a sensible *heating* process, which would require an *increase* in temperature, moving to the right on the chart rather than to the left.

SOLUTION 5

The U-factor for a building envelope component is the reciprocal of the total resistance.

$$U_{wall} = \frac{1}{R_{total}}$$

The total resistance of the wall, when all the constituent layers are arranged in series, is

$$R_{total} = R_{films} + R_{layers}$$

The total resistance of the air films is

$$\begin{aligned} R_{films} &= R_{outdoor} + R_{indoor} \\ &= 0.17 \ \frac{\text{hr-ft}^2\text{-°F}}{\text{Btu}} + 0.68 \ \frac{\text{hr-ft}^2\text{-°F}}{\text{Btu}} \\ &= 0.85 \text{ hr-ft}^2\text{-°F / Btu} \end{aligned}$$

The total resistance of the constituent layers making up the wall is the sum of the products of the unit resistance and thickness for each layer.

$$\begin{aligned} R_{\text{layers}} &= \sum tR_{\text{per in}} \\ &= (4.0 \text{ in})\left(0.10 \ \frac{\text{hr-ft}^2\text{-°F}}{\text{Btu-in}}\right) \\ &\quad +(2.0 \text{ in})\left(6.0 \ \frac{\text{hr-ft}^2\text{-°F}}{\text{Btu-in}}\right) \\ &\quad +(0.625 \text{ in})\left(0.90 \ \frac{\text{hr-ft}^2\text{-°F}}{\text{Btu-in}}\right) \\ &= 12.96 \text{ hr-ft}^2\text{-°F / Btu} \end{aligned}$$

The total resistance of the wall is

$$\begin{aligned} R_{\text{total}} &= 0.85 \ \frac{\text{hr-ft}^2\text{-°F}}{\text{Btu}} + 12.96 \ \frac{\text{hr-ft}^2\text{-°F}}{\text{Btu}} \\ &= 13.81 \text{ hr-ft}^2\text{-°F / Btu} \end{aligned}$$

The U-factor for the wall is

$$\begin{aligned} U_{\text{wall}} &= \frac{1}{13.81 \ \frac{\text{hr-ft}^2\text{-°F}}{\text{Btu}}} \\ &= 0.072 \text{ Btu/hr-ft}^2\text{-°F} \end{aligned}$$

The answer is (A).

Why Other Options Are Wrong

(B) This incorrect answer results from omitting the surface film resistances (convection).

(C) This incorrect answer results when the distinction between resistivity and resistance is not made and resistivities are not multiplied out by their corresponding thicknesses.

(D) This incorrect answer results from summing the reciprocals of each component resistance, rather than taking the reciprocal of the sum of all resistances.

SOLUTION 6

The relationship between pressure, volume, and temperature for an ideal gas is

$$\frac{p_1 V_1}{T_1} = \frac{p_2 V_2}{T_2}$$

Atmospheric pressure is

$$p_1 = 14.7 \text{ psia}$$

The absolute pressure after compression is

$$\begin{aligned} p_2 &= p_{\text{gage}} + p_{\text{atm}} \\ &= 100 \ \frac{\text{lbf}}{\text{in}^2} + 14.7 \ \frac{\text{lbf}}{\text{in}^2} \\ &= 114.7 \text{ psia} \end{aligned}$$

The air temperature in absolute units is

$$\begin{aligned} T_1 &= 70\text{°F} + 460\text{°} \\ &= 530\text{°R} \end{aligned}$$

The final temperature in absolute units of the gas following compression is

$$\begin{aligned} T_2 &= T_1\left(\frac{p_2 V_2}{p_1 V_1}\right) \\ &= (530\text{°R})\left(\frac{\left(114.7 \ \frac{\text{lbf}}{\text{in}^2}\right)(20\%)}{\left(14.7 \ \frac{\text{lbf}}{\text{in}^2}\right)(100\%)}\right) \\ &= 827\text{°R} \end{aligned}$$

Converting,

$$\begin{aligned} T_2 &= 827\text{°R} - 460\text{°} \\ &= 367\text{°F} \quad (370\text{°F}) \end{aligned}$$

The answer is (D).

Why Other Options Are Wrong

(A) This answer is wrong because the temperature was not converted to absolute units before the calculation.

(B) This answer is wrong because the pressure was not converted from gage to atmospheric.

(C) This answer is wrong because T_1 was multiplied by p_1V_1/p_2V_2 instead of by p_2V_2/p_1V_1 and then absolute temperature units (°R) were not converted back to standard units (°F).

SOLUTION 7

Heat removal from the chilled water cooling medium benefits from both subcooling and superheating. Thus, the net refrigeration effect can be determined by subtracting the enthalpy of the refrigerant entering the evaporator from the enthalpy of the refrigerant leaving the evaporator and entering the compressor. This process goes from the subcooled liquid state following expansion to the superheated vapor state just prior to compression.

The answer is (A).

Why Other Options Are Wrong

(B) This answer is wrong because subcooling has a net refrigerating effect on the chilled water medium.

(C) This answer is wrong because superheat has a net refrigerating effect on the chilled water medium.

(D) This answer is wrong because both superheat and subcooling have a net refrigerating effect on the chilled water medium.

SOLUTION 8

The coefficient of performance of a refrigeration cycle is

$$\begin{aligned}\text{COP} &= \frac{q_{\text{refrigeration}}}{W_{\text{compressor}}}\\ &= \frac{h_1 - h_4}{h_2 - h_1}\end{aligned}$$

From a table of R-123 properties, the enthalpy of the refrigerant after passing through the expansion valve, h_4, will be the same as that at the condensing temperature of 260°F, so the pressure is 198.3 psia.

$$h_4 = h_f = 75.7 \text{ Btu/lbm}$$

At the evaporating temperature of 100°F, the pressure is 20.8 psia, and the enthalpy is

$$h_g = 105 \text{ Btu/lbm}$$

Adding 40°F superheat to this point on an R-123 pressure-enthalpy diagram results in an enthalpy of approximately 112 Btu/lbm and an entropy of 0.20 Btu/lbm-°F. The net refrigeration is the difference between the superheated vapor enthalpy at 140°F, h_1, and the saturated liquid enthalpy after the expansion valve, h_4.

$$\begin{aligned}q_{\text{refrigeration}} &= h_1 - h_4\\ &= 112 \frac{\text{Btu}}{\text{lbm}} - 75.7 \frac{\text{Btu}}{\text{lbm}}\\ &= 36.3 \text{ Btu/lbm}\end{aligned}$$

Because this is an isentropic process, the compressor work is the difference between the vapor enthalpy at 20.8 psia and 140°F, h_1, and the vapor enthalpy at 198.3 psia, h_2, where both states have an entropy equal to 0.20 Btu/lbm-°F.

$$\begin{aligned}W_{\text{compressor}} &= h_2 - h_1\\ &= 129 \frac{\text{Btu}}{\text{lbm}} - 112 \frac{\text{Btu}}{\text{lbm}}\\ &= 17 \text{ Btu/lbm}\end{aligned}$$

$$\begin{aligned}\text{COP} &= \frac{q_{\text{refrigeration}}}{W_{\text{compressor}}}\\ &= \frac{36.3 \frac{\text{Btu}}{\text{lbm}}}{17 \frac{\text{Btu}}{\text{lbm}}}\\ &= 2.14 \quad (2.1)\end{aligned}$$

The answer is (C).

Why Other Options Are Wrong

(A) This incorrect answer results from not including the effect of superheat when determining the compressor input properties.

(B) This incorrect answer results from not including the effect of superheat when determining the refrigeration effect.

(D) This incorrect answer results from using the enthalpy at the *evaporating* saturation temperature, rather than the *condensing* saturation temperature, when calculating refrigeration effect.

SOLUTION 9

The entropy of a two-phase steam mixture is

$$s = s_f + x s_{fg}$$

Quality, x, is the ratio of the mass of vapor to the total mass.

$$\begin{aligned}x &= \frac{m_{\text{vapor}}}{m_{\text{total}}}\\ &= \frac{800 \text{ lbm} - 100 \text{ lbm}}{800 \text{ lbm}}\\ &= 0.875\end{aligned}$$

The pressure of the mixture is

$$\begin{aligned}p &= p_{\text{gage}} + p_{\text{atm}}\\ &= 20 \frac{\text{lbf}}{\text{in}^2} + 14.7 \frac{\text{lbf}}{\text{in}^2}\\ &= 34.7 \text{ psia}\end{aligned}$$

Interpolating between 30 and 40 psia on the saturated steam tables, for a value of 34.7 psia,

$$s_f = 0.3794 \text{ Btu/lbm-°R}$$

$$s_{fg} = 1.3094 \text{ Btu/lbm-°R}$$

The final entropy of the mixture is

$$\begin{aligned} s &= s_f + xs_{fg} \\ &= 0.3794\ \frac{\text{Btu}}{\text{lbm-°R}} + (0.875)\left(1.3094\ \frac{\text{Btu}}{\text{lbm-°R}}\right) \\ &= 1.525\ \text{Btu/lbm-°R} \quad (1.5\ \text{Btu/lbm-°R}) \end{aligned}$$

The answer is (B).

Why Other Options Are Wrong

(A) This answer is wrong because quality was miscalculated as the fraction of liquid, rather than vapor.

(C) This answer is wrong because the gage pressure was not converted to absolute.

(D) This answer is wrong because the quality term was left out of the equation, giving an answer equal to the entropy at its saturation vapor pressure.

SOLUTION 10

The only refrigeration cycle listed is the vapor compression cycle, which uses condensable vapor as the refrigerant.

The answer is (C).

Why Other Options Are Wrong

(A) This answer is wrong because the Rankine cycle is a power cycle used in steam engines.

(B) This answer is wrong because the Brayton cycle is a power cycle that uses gas as the working fluid.

(D) This answer is wrong because the Otto cycle is a combustion cycle for a four-stroke reciprocating engine.

SOLUTION 11

The condensation resistance factor, an industry-accepted measure of the condensation resistance of a fenestration system, is defined as

$$\text{CRF} = \frac{T_{\text{dp}} - T_{\text{cold}}}{T_{\text{hot}} - T_{\text{cold}}}$$

The dew point is determined from the psychrometric state point of the indoor air. From the psychrometric chart, the dew-point temperature at 74°F dry-bulb and 40% relative humidity is 48°F. The condensation resistance factor is

$$\begin{aligned} \text{CRF} &= \frac{48°\text{F} - (-12°\text{F})}{74°\text{F} - (-12°\text{F})} \\ &= 0.70 \end{aligned}$$

An industry reference for the condensation resistance factor is the ASHRAE Fundamentals Handbook. The equation used in this solution is Eq. 137 from Ch. 31, Fenestration.

The answer is (C).

Why Other Options Are Wrong

(A) This answer is the result of trying to subtract −12 from 48 and mistakenly getting 36.

(B) This incorrect answer is the result of subtracting 12 instead of −12 in both the numerator and denominator.

(D) This incorrect answer results from using the room wet-bulb temperature instead of the dew point.

SOLUTION 12

The Carnot efficiency for a power cycle operating between two constant temperature reservoirs is (with temperatures in absolute units)

$$\eta_{\text{Carnot}} = \frac{T_{\text{hot}} - T_{\text{cold}}}{T_{\text{hot}}}$$

The "hot" reservoir is the discharging condensate at 140°F, which is converted to absolute temperature.

$$\begin{aligned} T_{\text{hot}} &= 140°\text{F} + 460° \\ &= 600°\text{R} \end{aligned}$$

The "cold" reservoir is the 50°F sewer, which is also converted to absolute temperature.

$$\begin{aligned} T_{\text{cold}} &= 50°\text{F} + 460° \\ &= 510°\text{R} \end{aligned}$$

$$\begin{aligned} \eta_{\text{Carnot}} &= \frac{600°\text{R} - 510°\text{R}}{600°\text{R}} \\ &= 0.15 \quad (15\%) \end{aligned}$$

The answer is (A).

Why Other Options Are Wrong

(B) This answer results from dividing the cold temperature into the hot temperature in incorrect units.

(C) This incorrect answer results when temperatures are not converted to absolute units before calculating the Carnot efficiency.

(D) This answer results from incorrectly dividing the cold temperature into the hot temperature.

SOLUTION 13

Assuming full conversion of steam to water, the minimum required mass flow rate of steam is the rate of heating required divided by the latent energy.

$$\dot{m} = \frac{Q_{\text{heating}}}{h_{fg}}$$

The absolute pressure of the steam is the given gage pressure, plus atmospheric pressure.

$$\begin{aligned} p &= p_{\text{gage}} + p_{\text{atm}} \\ &= 5\ \frac{\text{lbf}}{\text{in}^2} + 14.7\ \frac{\text{lbf}}{\text{in}^2} \\ &= 19.7\ \text{psia} \end{aligned}$$

From interpolation in steam tables, the enthalpy change between saturated liquid and steam at 19.7 lbf/in^2 is

$$h_{fg} = 960.7\ \text{Btu/lbm}$$

The required steam mass flow rate is

$$\begin{aligned} \dot{m} &= \frac{47{,}000\ \frac{\text{Btu}}{\text{hr}}}{960.7\ \frac{\text{Btu}}{\text{lbm}}} \\ &= 48.9\ \text{lbm/hr} \quad (49\ \text{lbm/hr}) \end{aligned}$$

The answer is (D).

Why Other Options Are Wrong

(A) This answer results from erroneously using the enthalpy of saturated vapor (h_g) rather than the difference between the enthalpy of saturated vapor and saturated liquid (h_{fg}).

(B) This incorrect answer results from not converting the operating pressure of 5 psig to 19.7 psia.

(C) This incorrect answer results from using the enthalpy of evaporation at atmospheric pressure (970.3 Btu/lbm), which is a frequent reference temperature.

SOLUTION 14

The coefficient of performance is the refrigeration effect divided by the compressor work.

$$\text{COP} = \frac{Q}{P}$$

The power required to run the chiller compressor is

$$\begin{aligned} P &= \frac{Q}{\text{COP}} \\ &= \left(\frac{600\ \text{ton}}{5.5}\right)\left(\frac{12{,}000\ \frac{\text{Btu}}{\text{hr}}}{1\ \text{ton}}\right)\left(\frac{1\ \text{kW}}{3412\ \frac{\text{Btu}}{\text{hr}}}\right) \\ &= 384\ \text{kW} \end{aligned}$$

Operation cost is the product of the power, operation time, and energy rate. The cost of one hour is

$$\begin{aligned} C &= Ptr \\ &= (384\ \text{kW})(1\ \text{hr})\left(\frac{\$0.09}{\text{kW-hr}}\right) \\ &= \$34.60 \quad (\$35) \end{aligned}$$

The answer is (B).

Why Other Options Are Wrong

(A) This incorrect answer results from failure to convert tons to kilowatts.

(C) This incorrect answer results when the coefficient of performance is not taken into account.

(D) This answer results from a very common error when working with utility rates, neglecting a zero after the decimal place—in this case, reading \$0.09 as \$0.9.

SOLUTION 15

Aeration is the process of exposing a substance to air circulation. The others result in moisture removal.

The answer is (D).

Why Other Options Are Wrong

(A) Solid sorption is a dehumidification process. Air is passed through granular desiccant which has a vapor pressure below that of the humid air; this drives the water vapor from the air to the desiccant.

(B) Liquid desiccant is used in dehumidification. Air is brought in contact with liquid desiccant solution whose water vapor pressure can be controlled by temperature and concentration.

(C) Compressing air reduces its ability to hold moisture.

SOLUTION 16

It is common practice to determine the required supply air quantities based on sensible load only. Latent loads are brought into the analysis only when the cooling coil load is being determined. The sensible load from each source is

$$Q_{s,\text{solar}} = 50{,}000\ \text{Btu/hr}$$

$$Q_{s,\text{lighting}} = (20{,}000\ \text{W})\left(\frac{3.412\ \frac{\text{Btu}}{\text{hr}}}{1\ \text{W}}\right)$$

$$= 68{,}240\ \text{Btu/hr}$$

$$Q_{s,\text{people}} = 30{,}000\ \text{Btu/hr}$$

$$Q_{s,\text{steam}} = 0\ \text{Btu/hr}$$

There is no sensible component to the heat load from the steam tables. This load is of latent heat only.

The total sensible load is the sum of the individual loads.

$$Q_{s,\text{total}} = 148{,}240\ \text{Btu/hr} \quad (150{,}000\ \text{Btu/hr})$$

The answer is (B).

Why Other Options Are Wrong

(A) This answer is wrong because the lighting load, given in watts, was not converted to Btu/hr.

(C) This answer is wrong because the latent load due to people is included.

(D) This answer is wrong because the latent load due to the steam tables is included.

SOLUTION 17

Determine the apparatus dew-point of 40°F by plotting the dry-bulb temperature at saturation. This state point represents the condition of the air that actually comes in contact with the coil. Air that "bypasses" the coil fins and tubes remains at its original state point of 80°F dry-bulb and 65°F wet-bulb. The bypass factor indicates what fraction of air does not come into contact with the coil.

After plotting the points, draw a straight line connecting the state point of the air entering the coil with the apparatus dew-point (coil) temperature. A bypass factor of 0.20 indicates a coil with an efficiency of 0.80. Graphically, mark the distance that indicates a process travel of 80% of the distance from the conditions entering the coil to the coil temperature (apparatus dew point). This will be the condition leaving the coil.

The process is shown here with only the relevant lines of the psychrometric chart given.

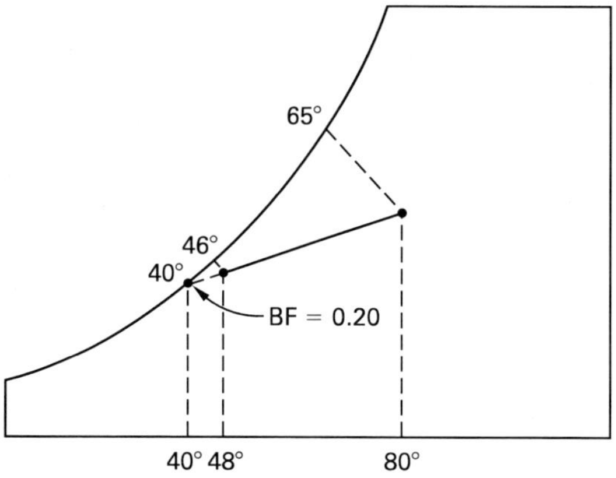

The answer is (B).

Why Other Options Are Wrong

(A) This incorrect answer results from assuming that the air mixture will leave the coil at the coil temperature (saturation). However, this is only true for the air that actually makes contact with the coil itself.

(C) This answer is incorrectly determined by drawing a straight line to the entering air dew point, rather than to the apparatus dew point.

(D) This incorrect answer comes from applying the bypass factor backward. The smaller the bypass factor, the closer to the apparatus dew point the conditioned air should be.

SOLUTION 18

An absorption refrigeration system operates directly from heat energy, rather than from a vapor compressor. The refrigerant is evaporated by the load normally, but is then absorbed by an absorbent liquid medium. This liquid medium is subsequently pressurized by heat energy, separated from the solution, and finally condensed back to its original state by a condenser and expansion valve.

The answer is (D).

Why Other Options Are Wrong

(A) Adsorption is the collection of gases, liquids, or solids to a surface, which is not part of this process.

(B) This answer is incorrect. A cascade refrigeration system is one with more than one refrigeration circuit.

(C) This answer is incorrect. Direct expansion refrigeration occurs when the cooling comes directly from the refrigerant, rather than, as here, from a secondary medium such as chilled water.

SOLUTION 19

The differential pressure across the doorway is proportional to the difference in density between two air columns 350 ft high, one at 10°F and the other at 70°F. Pressure due to air higher than the height of the building can be ignored.

When the inside air is warmer than the outside air, then the base of the building is depressurized relative to the outside, and the top is pressurized, because of the difference in weight between the two columns of air. Somewhere in between, often at midheight, is an elevation where the inside and outside pressures are equal. This is called the neutral plane (NPL).

The differential pressure across the envelope at any elevation is directly proportional to the distance between that elevation and the neutral plane.

$$\Delta p = \frac{\Delta \rho g (h - h_{\text{NPL}})}{g_c}$$

From the psychrometric chart, the specific volume of dry inside air at 70°F is 13.3 ft^3/lbm. The air density is the reciprocal of the specific volume.

$$\begin{aligned} \rho_i &= \frac{1}{v_i} \\ &= \frac{1}{13.3\ \frac{\text{ft}^3}{\text{lbm}}} \\ &= 0.075\ \text{lbm/ft}^3 \end{aligned}$$

Similarly, the density of dry outside air at 10°F is

$$\begin{aligned} \rho_o &= \frac{1}{v_o} \\ &= \frac{1}{11.8\ \frac{\text{ft}^3}{\text{lbm}}} \\ &= 0.085\ \text{lbm/ft}^3 \end{aligned}$$

Assuming the neutral plane is at midheight,

$$\begin{aligned} h_{\text{NPL}} &= \frac{350\ \text{ft}}{2} \\ &= 175\ \text{ft} \end{aligned}$$

The pressure differential across the doorway is

$$\begin{aligned} \Delta p &= \frac{\Delta \rho g (h - h_{\text{NPL}})}{g_c} \\ &= \frac{\left(0.085\ \frac{\text{lbm}}{\text{ft}^3} - 0.075\ \frac{\text{lbm}}{\text{ft}^3}\right) \times \left(32.2\ \frac{\text{ft}}{\text{sec}^2}\right)(0\ \text{ft} - 175\ \text{ft})}{32.2\ \frac{\text{ft-lbm}}{\text{lbf-sec}^2}} \\ &\quad \times \left(\frac{1\ \text{ft}}{12\ \text{in}}\right)^2 \left(\frac{27.7\ \text{in wg}}{1\ \frac{\text{lbf}}{\text{in}^2}}\right) \\ &= -0.0337\ \text{in wg} \quad (-0.34\ \text{in wg}) \end{aligned}$$

The answer is (A).

Why Other Options Are Wrong

(B) This incorrect answer results when the conversion from pounds per square foot to inches of water gage is omitted.

(C) This incorrect answer results when the term g_c is omitted.

(D) This incorrect answer results when the specific volumes are not converted to densities before they are inserted in the equation.

SOLUTION 20

The absolute pressure in the bladder of the expansion tank will add to the hydrostatic pressure at the bottom of the tank. The total pressure at the bottom of the tank is the sum of the two.

$$\begin{aligned} p_{\text{total}} &= p_{\text{abs}} + p_{\text{hyd}} = p_{\text{gage}} + p_{\text{atm}} + \frac{\rho g h}{g_c} \\ &= 15\ \frac{\text{lbf}}{\text{in}^2} + 14.7\ \frac{\text{lbf}}{\text{in}^2} \\ &\quad + \left(\frac{\left(62.34\ \frac{\text{lbm}}{\text{ft}^3}\right)\left(32.2\ \frac{\text{ft}}{\text{sec}^2}\right)(2\ \text{ft})}{32.2\ \frac{\text{ft-lbm}}{\text{lbf-sec}^2}}\right) \\ &\quad \times \left(\frac{1\ \text{ft}}{12\ \text{in}}\right)^2 \\ &= 30.57\ \text{psia} \quad (31\ \text{psia}) \end{aligned}$$

The answer is (B).

Why Other Options Are Wrong

(A) This answer is wrong because the tank gage pressure was not converted to absolute pressure.

(C) This answer is wrong because the unit conversion g_c was left out of the calculation.

(D) This answer is wrong because the static head portion was not converted from square feet to square inches.

2 Depth

PRINCIPLES

PROBLEM 1

Cool supply air is introduced into a room at 56°F dry-bulb and 54°F wet-bulb temperature. At the same time, the room experiences a sensible cooling load of 100,000 Btu/hr and a latent cooling load of 45,000 Btu/hr. The room thermostat maintains the space dry-bulb temperature at 75°F. Assuming that the room is at about sea level, the resulting humidity ratio will be

(A) 0.0011 lbm/lbm

(B) 0.0079 lbm/lbm

(C) 0.010 lbm/lbm

(D) 0.017 lbm/lbm

Hint: The room thermostat can only control based on dry-bulb temperature. The final humidity ratio will therefore be found where the sensible/latent process line intercepts the dry-bulb temperature line.

PROBLEM 2

Conditioned air at 55°F dry-bulb and 54°F wet-bulb temperature is distributed throughout a theater that seats 300. The system is controlled so that the air is returned to the fan system at 75°F. The lighting gain to the space is 10 kW, and the heat gain from each person is 200 Btu/hr sensible and 150 Btu/hr latent. The relative humidity of the air returning to the fan system is most nearly

(A) 50%

(B) 60%

(C) 80%

(D) 90%

Hint: The sensible heat ratio plotted on the psychrometric chart will show the relationship between the entering supply air and the final state point.

PROBLEM 3

A room is to be conditioned to 76°F dry-bulb temperature and 50% relative humidity with 55°F supply air from the system shown.

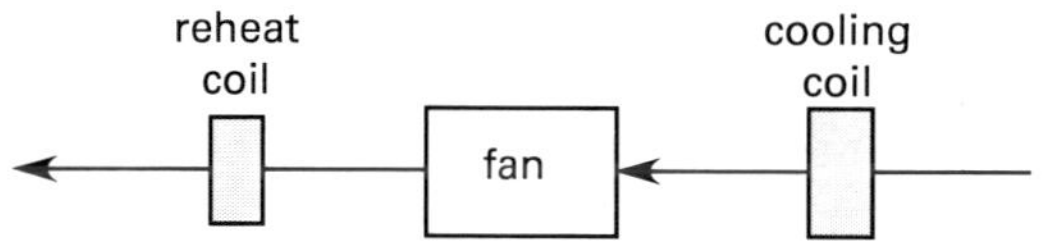

Air leaves the cooling coil at a rate of 2000 ft^3/min at 44°F dry-bulb and 43°F wet-bulb temperature. The temperature rise across the fan is 2°F. The reheat coil load is most nearly

(A) 19,800 Btu/hr

(B) 20,600 Btu/hr

(C) 24,600 Btu/hr

(D) 69,200 Btu/hr

Hint: On a standard psychrometric chart, the psychrometric process line must first be drawn to the cooling coil leaving condition. Then the reheat process line can be drawn to the required supply air temperature.

PROBLEM 4

A refrigeration system to be installed must be capable of cooling 10,000 lbm of pork shoulder from 40°F to 10°F in 2 hr. The pork shoulder has a freezing point of 28°F, water content of 73%, a specific heat above freezing of 0.86 Btu/lbm-°F and below freezing of 0.53 Btu/lbm-°F, and a thermal conductivity of 0.71 Btu/hr-ft-°F. Assuming atmospheric conditions, the capacity the refrigeration system must have is most nearly

(A) 625,000 Btu/hr

(B) 635,000 Btu/hr

(C) 820,000 Btu/hr

(D) 1,250,000 Btu/hr

Hint: The required sensible cooling will vary depending upon whether the pork is above or below freezing, and the latent heat removal will depend upon the water content of the pork.

Depth

PROBLEM 5

A refrigeration system has an energy efficiency ratio (EER) of 7.5 Btu/W-hr. The system provides 1500 ft^3/min of supply air through a cooling coil. An electric reheat coil is to be added to the supply air discharge, and the modified system will be used to process air from 75°F dry-bulb/60% relative humidity to 75°F dry-bulb/50% relative humidity. Each coil has an efficiency of very nearly 100%. The minimum electric power needed is

(A) 6100 W

(B) 9800 W

(C) 16,000 W

(D) 23,000 W

Hint: It is not possible to go directly to the desired psychrometric state point. The air must be first brought to the desired humidity ratio and then heated to the desired temperature in two steps.

PROBLEM 6

A hospital operating room with a 9 ft ceiling and floor dimensions of 20 ft by 25 ft must be provided with a minimum of eight air changes per hour of outdoor air at 10°F dry-bulb and 8°F wet-bulb temperature. Indoor conditions of 70°F dry-bulb and 40% relative humidity must be maintained. The humidifier for the system must be sized to supply humidification at a rate no less than approximately

(A) 4.9 lbm/hr

(B) 15 lbm/hr

(C) 16 lbm/hr

(D) 870 lbm/hr

Hint: The flow rate will be equal to the product of the air volume of the hospital room and the number of air changes per hour.

PROBLEM 7

A stream of air flows through a humidifier at a rate of 4000 ft^3/min, at a dry-bulb temperature of 70°F and a relative humidity of 20%. The air is humidified with 15 psig saturated steam to a dew point of 50°F. The temperature of the air downstream of the humidifier is most nearly

(A) 50°F

(B) 68°F

(C) 70°F

(D) 72°F

Hint: On the psychrometric chart, the final state of the airstream lies on the straight line that passes through the initial conditions and whose slope $\Delta h/\Delta\omega$ is set by the specific enthalpy of the injected moisture.

PROBLEM 8

The room enclosing an indoor swimming pool is to be maintained at 82°F and 60% relative humidity. The evaporation rate of the 80°F water is 350 lbm/hr. Dehumidification will be accomplished by the introduction of outdoor air. The outdoor design conditions are 90°F dry-bulb and 68°F wet-bulb temperature. The amount of outdoor air required to dehumidify the pool area on a "design day" is most nearly

(A) 1360 ft^3/min

(B) 16,800 ft^3/min

(C) 18,100 ft^3/min

(D) 1.09×10^4 ft^3/min

Hint: To dehumidify with outside air, the air quantity must be sufficient to remove the pool evaporation using the difference between the indoor and outdoor humidity ratios.

PROBLEM 9

A humidifier is to be installed in the 10,000 ft^3/min air system serving the new wing of a hospital. The system can provide a variable quantity of outside air, but should be capable of maintaining space conditions of 75°F and 40% relative humidity when 4000 ft^3/min outdoor air at 10°F dry-bulb and 8°F wet-bulb temperature is mixed with 6000 ft^3/min recirculated air and preheated to 55°F. The required humidifier capacity is most nearly

(A) 1.5 lbm/hr

(B) 8.2 lbm/hr

(C) 36 lbm/hr

(D) 120 lbm/hr

Hint: The humidification load can be calculated as $H = 60\rho\dot{V}(\omega_i - \omega_o)((t_i - t_m)/(t_i - t_o))$.

PROBLEM 10

The pump shown delivers 46 ft total dynamic head.

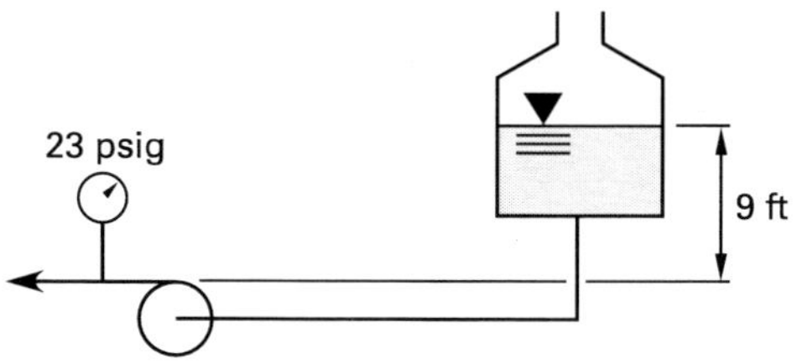

Friction losses between the tank and the pump can be assumed to be negligible. The estimated velocity through the pipe is most nearly

(A) 1.4 ft/sec

(B) 11 ft/sec

(C) 59 ft/sec

(D) 120 ft/sec

Hint: The pressure gauge is capable of sensing static head, but not velocity head.

PROBLEM 11

Pressure drop caused by friction in fully developed fluid flow can be determined directly by using

(A) the first law of thermodynamics

(B) the Darcy-Weisbach equation

(C) the Bernoulli equation

(D) computational fluid dynamics

Hint: The Moody friction factor is featured prominently in this solution.

PROBLEM 12

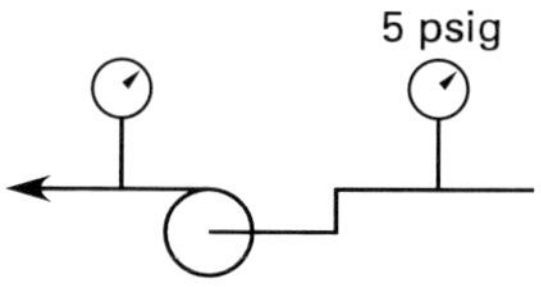

A pump is selected to provide 300 gpm of water against a total dynamic head of 60 ft wg. It is installed in a 3 in, schedule-40 pipe with pressure gauges at the inlet and outlet. When the pump is started up, the gauge on the inlet side reads 5 psi. Assume that friction and fitting losses are negligible for the short distance. The reading of the discharge pressure gauge will be most nearly

(A) 31 psig

(B) 32 psig

(C) 46 psig

(D) 65 psig

Hint: Try rearranging the Bernoulli equation to solve for static head at the outlet. Not all the information given is needed.

PROBLEM 13

A baseboard heater is to be purchased for a bedroom addition. The heater must have the capacity to maintain the room at 70°F against outdoor temperatures as low as 15°F. The surfaces exposed to the outdoors are as follows.

surface	area (ft^2)	U-factor $\left(\frac{\text{Btu}}{\text{ft}^2\text{-hr-°F}}\right)$
wall	120	0.094
window	24	0.65
ceiling	200	0.037

The available electric baseboard heaters have rated capacities of 250 W per foot of baseboard length. The minimum number of feet of wall-mounted baseboard heater required is

(A) 2 ft

(B) 3 ft

(C) 7 ft

(D) 8 ft

Hint: The heat loss for a surface exposed to indoor and outdoor ambient conditions is the product of the U-factor, the surface area, and the temperature difference across the entire component.

PROBLEM 14

An outdoor solarium is to be cooled with 58°F supply air. The solarium contains 1500 ft^2 of glazing with a shading coefficient of 0.81. An infiltration latent load of 35,000 Btu/hr is also imposed upon the space. The solar cooling load factor for the glazing at a given hour is

Depth

193 Btu/hr-ft^2. The quantity of supply air required to maintain the space temperature of 74°F is most nearly

(A) 13,600 ft^3/min

(B) 15,600 ft^3/min

(C) 16,800 ft^3/min

(D) 815,000 ft^3/min

Hint: The cooling load through glazing is a product of the solar cooling load factor and the glazing area, if single-pane clear glass is used. For multiple panes or tinted glazing, the shading coefficient must be employed as well.

PROBLEM 15

An office worker sits 5 ft away from a window 22 ft long and 6.5 ft high. The window surface temperature is 40°F and all other interior surfaces are 70°F.

In the angle factor graph shown, a is the window's length, b is its height, and c is the distance to the worker.

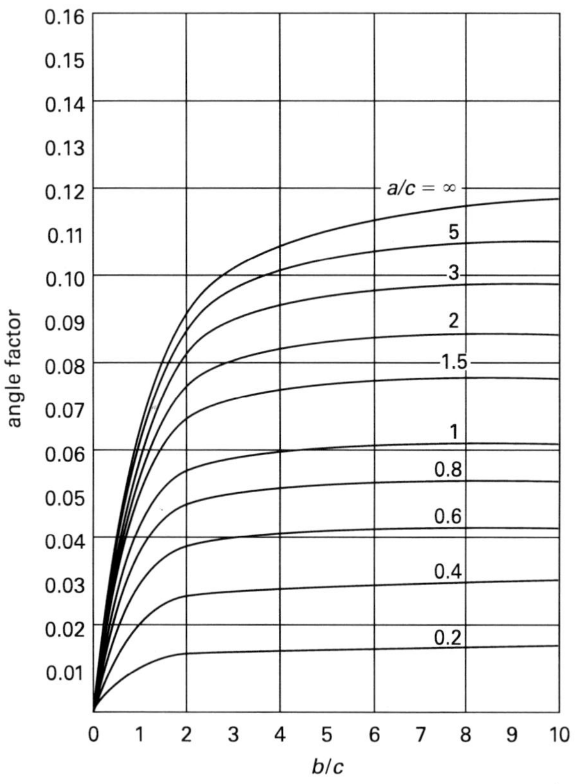

Adapted from *ASHRAE Standard 55*.

The mean radiant temperature experienced by the worker is most nearly

(A) 40°F

(B) 55°F

(C) 68°F

(D) 70°F

Hint: The angle factor for the window can be determined from the graph using the ratios of the given parameters. The total of all angle factors must be unity.

PROBLEM 16

A heat exchanger must be selected to offset the evaporation losses of a public indoor swimming pool. The pool measures 165 ft by 45 ft. The air in the building is kept at 82°F and 55% relative humidity. The pool water is kept at 80°F, and the air velocity will be kept to 25 ft/min. The evaporation rate of the pool is most nearly

(A) 52 lbm/hr

(B) 160 lbm/hr

(C) 320 lbm/hr

(D) 380 lbm/hr

Hint: The evaporation rate is proportional to the difference between the saturation vapor pressure taken at the surface water temperature and the saturation pressure taken at the room air dew point. This mass transfer problem is solved in the industry using well-established empirical relationships, such as those found in the ASHRAE Applications Handbook.

PROBLEM 17

An evaporative cooler uses a 53°F water spray to precool 8000 ft^3/min of outdoor air, from 96°F dry-bulb and
55°F wet-bulb temperature to 62°F dry-bulb. The saturation efficiency of the process is most nearly

(A) 21%

(B) 80%

(C) 83%

(D) 100%

Hint: The theoretical dry-bulb temperature achievable is the wet-bulb temperature. The saturation efficiency of the process is the fraction of the difference between entering dry-bulb temperature and the outdoor wet-bulb temperature.

PROBLEM 18

A 104 gpm recirculating domestic hot water system serves a dormitory, which experiences a load of 44 gpm. Pipe losses are 100,000 Btu/hr for the water supplied to the dorm and 82,000 Btu/hr for the water returning to the water heater. The water leaving the heater is 125°F, and the makeup water is 53°F. The temperature of the water entering the heater is most nearly

(A) 91.8°F

(B) 92.5°F

(C) 93.2°F

(D) 159°F

Hint: The first-law analysis of the heat loss, mass flow, and specific heat will yield the temperature of the water to and from its destination. The water at return conditions (both flow and temperature) can then be combined with the makeup water as a traditional mixing problem to determine the final temperature.

PROBLEM 19

Skylights to be installed in a Fairbanks office building are specified with frames having an R-value of 2.0 hr-ft^2-°F/Btu, not including indoor and outdoor film coefficients of 1.60 Btu/hr-ft^2-°F and 6.0 Btu/hr-ft^2-°F, respectively. The outdoor design temperature is −60°F. The maximum relative humidity that can be maintained in the 72°F heated space without experiencing condensation on the frames is most nearly

(A) 0%

(B) 20%

(C) 35%

(D) 100%

Hint: The maximum relative humidity at room temperature is limited by the dew-point temperature equal to the surface temperature of the skylight frame.

PROBLEM 20

Air enters a cooling tower at 50°F dry-bulb and 49°F wet-bulb temperature. It leaves the tower saturated at 70°F. Which condition is a likely outcome of this?

(A) adsorption

(B) condensation

(C) fog

(D) sublimation

Hint: Draw the process line between the two state points on the psychrometric chart.

PROBLEM 21

A natural draft cooling tower has a 400 ton heat rejection load. The tower operates at an efficiency of 52% when the outdoor wet-bulb temperature is 75°F and the entering water is 90°F. The temperature of the water leaving the tower under these conditions is most nearly

(A) 67°F

(B) 75°F

(C) 82°F

(D) 83°F

Hint: The efficiency of an evaporative cooling tower is defined by both the *range* and the *approach*.

PROBLEM 22

The parameter that has the greatest limiting effect on the thermal performance of an open, evaporative cooling tower is

(A) approach

(B) range

(C) dry-bulb temperature

(D) wet-bulb temperature

Hint: Consider whether sensible or latent performance is more significant.

PROBLEM 23

Air at 80°F dry-bulb and 67°F wet-bulb temperature enters a 42°F cooling coil and leaves at 53°F dry-bulb. The air is then reheated to 66°F. The cooling coil efficiency is most nearly

(A) 0.3

(B) 0.4

(C) 0.7

(D) 1

Hint: The coil efficiency accounts for that fraction of the air that makes contact with the coil and undergoes the psychrometric process terminating at the coil temperature.

PROBLEM 24

A four-row cooling coil can reduce the temperature of entering air from 85°F to 54°F when the coil (apparatus dew-point) temperature is 42°F and incoming chilled water is at 38°F. When this same coil meets entering air at 80°F with an apparatus dew point of 38°F, the discharge air temperature will be most nearly

(A) 45°F

(B) 49°F

(C) 50°F

(D) 52°F

Hint: Determining the bypass factor will allow estimation of temperatures at other conditions.

PROBLEM 25

A control action where the controller output varies infinitely over the range of the controller is classified as

(A) two-position action

(B) floating action

(C) modulating control

(D) fuzzy logic

Hint: Another term used for this kind of control is "proportional."

PROBLEM 26

Refrigeration condensers are rated based upon their ability to reject the total heat that comes from

(A) subcooling

(B) compressor work

(C) compressor work and net refrigeration effect

(D) superheating

Hint: The building or process is not the only source of heat to consider.

PROBLEM 27

After the control valve is activated, it takes 50 sec for the discharge air of a heating coil to achieve a new steady-state value. The time constant for the system is most nearly

(A) 25 sec

(B) 32 sec

(C) 50 sec

(D) 60 sec

Hint: Review the ASHRAE definition of the time constant for a system.

PROBLEM 28

A fan that has served an auditorium for a number of years has these characteristics.

airflow	35,000 ft^3/min
total pressure	2.0 in wg
speed	1050 rpm
brake horsepower	16.2 hp
motor horsepower	20.0 hp

The building facilities manager wishes to increase system airflow to the maximum extent that the existing fan motor will allow. The maximum achievable airflow is most nearly

(A) 33,000 ft^3/min

(B) 38,000 ft^3/min

(C) 39,000 ft^3/min

(D) 43,000 ft^3/min

Hint: Apply the third fan law.

PROBLEM 29

A pitot tube connected to a manometer is inserted into a supply-air duct facing upstream so that one side of the manometer reads total pressure and the other side reads static pressure. The air velocity is known to be 2500 ft/min. The distance, in inches, between the two water levels in the manometer is most nearly

(A) 0.033 in

(B) 0.39 in

(C) 27.0 in

(D) 1400 in

Hint: Total pressure is the sum of velocity pressure and static pressure in an airstream. The difference between the two columns of water gives velocity pressure, in inches. This reading can be predicted by evaluation of the velocity head term in the Bernoulli equation.

APPLICATIONS

PROBLEM 30

A design specification requires that a centrifugal pump deliver 380 gpm of water against 115 ft total dynamic head at 1750 rpm. A manufacturer makes a pump with a 12.5 in diameter impeller. The manufacturer's pump curve indicates that the pump can deliver the same flow at the same speed against 140 ft of total dynamic head.

To comply with the specification, the manufacturer will have to trim the pump impeller to a diameter of most nearly

(A) 7.0 in

(B) 8.4 in

(C) 10 in

(D) 11 in

Hint: The modified pump may be considered homologous to the pump for which data is available. As a result, the similarity laws apply.

PROBLEM 31

The existing 2000 ton chiller at a large hospital must be replaced. The type most likely to be installed is a

(A) reciprocating chiller

(B) scroll chiller

(C) centrifugal chiller

(D) screw chiller

Hint: Chiller selection is based in part on capacity requirements. Only one of the listed machines is available with the needed capacity.

PROBLEM 32

A direct-contact deaerator tank operates at 5 psig. The tank handles 1000 lbm/hr of 160°F condensate returned from a boiler system. The amount of 350°F steam at 30 psig required to drive the oxygen out of the condensate return is most nearly

(A) 2.0 lbm/hr

(B) 66 lbm/hr

(C) 68 lbm/hr

(D) 91 lbm/hr

Hint: The amount of steam required is that necessary to heat the condenser water to its saturation temperature at 5 psig.

PROBLEM 33

A steam boiler has a rating of 100 boiler horsepower. The boiler is operated at 15 psig with water entering at 180°F. The maximum rate of steam that can be generated is most nearly

(A) 3300 lbm/hr

(B) 3340 lbm/hr

(C) 3350 lbm/hr

(D) 3540 lbm/hr

Hint: A boiler horsepower is defined as 33,475 Btu/hr.

PROBLEM 34

A new central furnace has a steady-state efficiency of 81% and operates with a positive vent pressure. For classification purposes, ANSI Standard Z21.47/CSA 2.3 will categorize the furnace as

(A) Category I—a central furnace that operates with a non-positive vent pressure and a flue loss no less than 17%

(B) Category II—a central furnace that operates with a non-positive vent pressure and a flue loss less than 17%

(C) Category III—a central furnace that operates with a positive vent pressure and a flue loss no less than 17%

(D) Category IV—a central furnace that operates with a positive vent pressure and a flue loss less than 17%

Hint: Note that a central furnace with a positive vent pressure can only be classified in two of the categories.

PROBLEM 35

An air system serves a wing of a building. The chilled water coil for the system requires that, at design conditions, entering water is at 42°F and leaving water at 54°F The total cooling capacity of the coil is 650,000 Btu/hr and the total dynamic head of the water just upstream of the coil is 60 ft wg. The specifications require that the cooling coil control valve pressure drop be no greater than 25% of the total pressure at wide-open condition. The required control valve flow coefficient is most nearly

(A) 17

(B) 28

(C) 42

(D) 2500

Hint: The required flow coefficient will be given by $C_v = Q_{gpm} / \sqrt{\Delta p_{psi} / SG}$.

Depth

PROBLEM 36

A centrifugal pump with a 100 hp motor operates at 1750 rpm. It requires 86.6 brake horsepower to deliver 1400 gal/min of water at a total dynamic head of 147 ft wg. If the flow is reduced to 1050 gal/min, the motor can be changed out to one with a lower installed horsepower. The lowest acceptable standard motor size would be

(A) 35 hp

(B) 40 hp

(C) 50 hp

(D) 75 hp

Hint: The affinity laws are needed to solve this problem.

PROBLEM 37

The reversing valve in a unitary air conditioner is

(A) a device in a three-pipe fan coil to switch between hot water and chilled water

(B) a two-way valve used to reverse the water flow direction through a coil

(C) a device in a heat pump to change between heating and cooling cycles

(D) a pneumatic control system component that reverses the air signal

Hint: The reversing valve is associated with a refrigeration process.

PROBLEM 38

In the design of a refrigeration system, care should be taken to pitch the compressor discharge piping downward from the compressor, in order to

(A) provide sufficient refrigerant head going into the condenser

(B) ensure that the refrigerant can develop the minimum velocity needed to enter the condenser

(C) save energy by reducing the elevation change necessary

(D) prevent oil from draining back into the compressor when it is not operating

Hint: Two of these choices would in fact give no advantage to the system, and a third would be of insignificant benefit.

PROBLEM 39

When steel pipe is designated as "schedule 40," the schedule number refers to the

(A) maximum allowable continuous velocity needed to avoid excessive erosion

(B) ratio of service pressure to allowable stress

(C) absolute interior pipe roughness, in thousandths of an inch

(D) Brinell hardness number of the pipe

Hint: The designating number is dimensionless.

PROBLEM 40

A centrifugal pump is located at 7000 ft above sea level, where the atmospheric pressure is 11.3 psia. The pump is required to deliver 90°F water from a storage reservoir 6 ft below the pump's centerline. Friction and fitting losses are 8.0 ft wg. The pump curve indicates a net positive suction head requirement (NPSHR) of 9 ft wg. The net positive suction head available for the pump is most nearly

(A) – 3.4 ft

(B) 10 ft

(C) 12 ft

(D) 16 ft

Hint: The net positive suction head available (NPSHA) is determined by subtracting any reductions in head from the local atmospheric pressure available and adjusting as well for the vapor pressure of the water.

PROBLEM 41

The minimum energy performance requirements of various building HVAC system components are affected by all the following standards but one. Which one is the exception?

(A) ANSI/ASHRAE/IESNA Standard 90.1

(B) ASHRAE Standard 55

(C) ANSI Z21.10/CSA 4

(D) CTI Standard 201

Hint: The standard dealing with human comfort does not directly impact energy *performance* requirements for building systems.

PROBLEM 42

It is necessary to ventilate the vault for a 5 MW (5×10^6 W) transformer, located in the California desert, which has an outdoor design temperature of 105°F. The transformer, which is 97% efficient, serves a nearby facility that is conditioned to 74°F. The transformer vault must be maintained at a temperature no higher than 135°F. The amount of outdoor air that must be circulated through the vault is most nearly

(A) 4600 ft^3/min

(B) 7800 ft^3/min

(C) 16,000 ft^3/min

(D) 530,000 ft^3/min

Hint: The transformer losses represent the heat that must be continuously removed from the vault.

PROBLEM 43

According to the Uniform Mechanical Code, the allowable sheet metal gage for rectangular ductwork

(A) increases with increasing duct static pressure requirements

(B) increases with increasing dimension of the duct's longest side

(C) decreases with increasing dimension of the duct's longest side

(D) decreases with increasing dimension of the duct's shortest side

Hint: Sheet metal gage is lower for correspondingly greater sheet metal thicknesses.

PROBLEM 44

Of these refrigerant classifications, the greatest flammability is represented by

(A) A1

(B) A3

(C) B2

(D) C2

Hint: The letter designations refer to toxicity.

PROBLEM 45

ASHRAE Standard 55 defines human comfort in the HVAC industry. It uses the "comfort chart" to define the combination of factors that can be used to quantify human comfort for most of a building's occupants. The two properties used in this evaluation are humidity and

(A) dry-bulb temperature

(B) operative temperature

(C) mean radiant temperature

(D) air velocity

Hint: The parameter accounts for both convection and radiation effects.

PROBLEM 46

An HVAC system is to be installed in a 12,000 ft^2 research facility serving 85 people. The HVAC system provides 17,000 cfm of total supply air. Under the ventilation rate procedure described in ANSI/ASHRAE Standard 62.1, the minimum permissible rate of outside air is most nearly

(A) 1100 cfm

(B) 1800 cfm

(C) 2600 cfm

(D) 17,000 cfm

Hint: ASHRAE Standard 62.1 allows for the occupied floor area as well as the number of occupants to be factored into the required outside air rate.

PROBLEM 47

A 11,500 ft^2 supermarket receives 12,000 ft^3/min of conditioned air. The outside-air damper is set to the minimum that will comply with ANSI/ASHRAE Standard 62.1. The percentage of outside air is most nearly

(A) 10%

(B) 12%

(C) 25%

(D) 33%

Hint: For a supermarket, ASHRAE Standard 62.1 default occupancy density is 8 people per 1000 ft^2 and the default combined outdoor air rate is 7.5 ft^3/min per person plus 0.06 $ft^3/min/ft^2$.

PROBLEM 48

High sidewall grilles have been selected to deliver supply air to a high school multipurpose room that measures 30 ft across. The grilles will be installed in the wall on one side of the room and will blow air toward the other side. According to ASHRAE, the cooling load density of 20 Btu/hr-ft^2 results in a recommended throw ratio (T_{50}/L) of 1.5. To achieve good occupant comfort and a maximum air diffusion performance index (ADPI), the throw used in the grille selection should be most nearly

(A) 20 ft

(B) 23 ft

(C) 30 ft

(D) 45 ft

Hint: The characteristic dimension for a high sidewall grille is the distance, *L*, from the grille to the opposite wall.

PROBLEM 49

During a peer review of an air distribution system, an engineer is asked to determine if the diffuser was selected properly for the space in which it is located. The ceiling diffuser has the following published performance information.

neck velocity (fpm)	airflow (cfm)	throw (50 fpm)	throw (100 fpm)	throw (150 fpm)
450	310	5.0	4.7	4.2
500	360	6.5	5.5	4.5
550	405	7.0	6.0	5.0
600	462	8.0	7.0	6.0
650	505	9.0	8.0	7.0
700	557	10.0	9.0	8.0

The room sensible load is 10,000 Btu/hr, and the differential between design room temperature and supply air temperature is 20°F. The proper throw for diffuser selection is most nearly

(A) 5.0 ft

(B) 6.0 ft

(C) 7.0 ft

(D) 8.0 ft

Hint: The ASHRAE air diffusion performance index (ADPI) selection method is based on a terminal velocity of 50 ft/min (T_{50}).

PROBLEM 50

A computer room air conditioner (CRAC unit) is installed in a raised-floor data center environment. The unit discharges 40,000 ft³/min of conditioned air downward into a 1.5 ft deep raised-floor plenum, as shown. The air is distributed horizontally under the floor in a 360° pattern with an initial velocity of 25 ft/sec and a static pressure of 0.75 in wg.

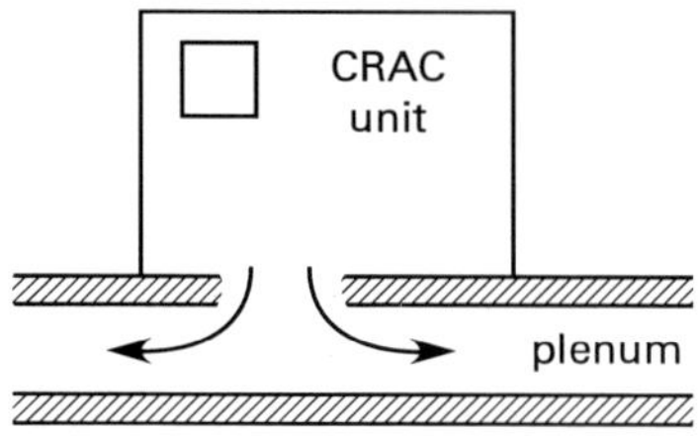

Assuming no static pressure losses occur in the plenum, the static pressure in inches of water gage at the first open discharge tile 8 ft from the unit is most nearly

(A) 0.61 in wg

(B) 0.75 in wg

(C) 0.87 in wg

(D) 0.89 in wg

Hint: The static pressure in the plenum will increase as the velocity pressure decreases.

PROBLEM 51

For purposes of friction loss calculation, the equivalent diameter of a rectangular duct measuring 24 in by 36 in is most nearly

(A) 16 in

(B) 29 in

(C) 32 in

(D) 33 in

Hint: The equivalent diameter is the diameter of the circular duct that would have the same fluid resistance and capacity as a given rectangular duct. It is not the same as hydraulic diameter.

PROBLEM 52

An engineer is designing a supply duct distribution system using the equal friction method. This approach will result in ductwork in which

(A) friction losses are equal to velocity pressure reductions

(B) friction losses are equal based on economic optimization

(C) friction losses per unit length are equal throughout the entire duct

(D) friction losses are equal between parallel branches in the system

Hint: The "equal" in equal friction means duct sizing decisions are based on maintaining the same losses throughout the system.

PROBLEM 53

Conditioned air is discharged from a 4 ft by 4 ft cooling coil at 500 ft/min. The air passes through a duct transition and changes course 90° by way of a 24 in by 24 in rectangular elbow. The elbow has a documented fitting coefficient of 1.4. The static pressure loss through the elbow, in inches of water gage, is most nearly

(A) 0.010 in wg

(B) 0.022 in wg

(C) 0.25 in wg

(D) 0.35 in wg

Hint: The loss of static pressure through the fitting is proportional to the velocity pressure. The fitting coefficient is the proportionality constant.

PROBLEM 54

A fan delivers 23,000 ft^3/min of airflow to a duct 40 in wide by 32 in high. A balancing damper is to be installed at an angle of 35° downstream of the fan to reduce the airflow. The following table shows the relationship between the duct fitting coefficient (C), duct shape (ratio of height to width, h/w), and damper angle (θ).

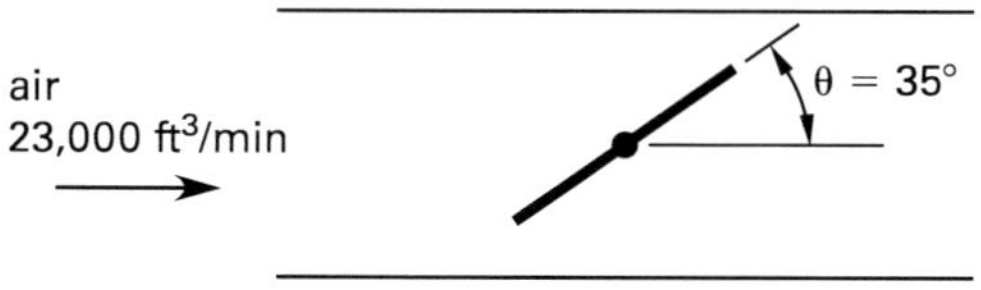

balancing valve fitting coefficient, C

θ	h/w			
	0.12	0.25	1.00	2.00
0°	0.04	0.08	0.08	0.13
10°	0.30	0.33	0.33	0.35
20°	1.10	1.18	1.18	1.25
30°	3.00	3.30	3.30	3.60
40°	8.00	9.00	9.00	10.0
50°	23.0	26.0	26.0	29.0
60°	60.0	70.0	70.0	80.0
65°	100	128	128	155
70°	190	210	210	230
90°	99999	99999	99999	99999

According to the illustration and table, the static pressure loss resulting from the newly installed balancing damper, in inches of water gage, is most nearly

(A) 0.060 in wg

(B) 0.21 in wg

(C) 0.42 in wg

(D) 2.6 in wg

Hint: The relationship between static pressure drop through any duct fitting and velocity pressure is $\Delta p_{\text{static}} = Cp_v$.

PROBLEM 55

A domestic hot water system runs 24 hr per day, 365 days per year, yet it is only needed during the daytime operation of an office building. Nameplate data give the information shown for the hot water circulation pump and its motor.

voltage	460 V
phases	3
amperage	8 A
efficiency	75%
total dynamic head	55 ft
flow	60 gpm
power factor	0.88

A time clock could be installed to shut off the pump 12 hr per day. The local utility charges $0.08/kW-hr and $16/kW for the maximum monthly peak demand. The potential annual savings that would result from the time clock is most nearly

(A) $1100/yr

(B) $2000/yr

(C) $2600/yr

(D) $3000/yr

Hint: The electricity rate is in kilowatt-hours, so it will be necessary to determine the instantaneous power draw (kW) and multiply that by the reduced hours of operation to determine energy consumption (kW-hr).

PROBLEM 56

A gas furnace that provides 1000 ft^3/min of supply air is selected to provide heating for a school classroom. The 1240 ft^2 room is designed for 7-year-old students. The furnace specification requires that the mixed air entering the unit be introduced to the unit at a temperature no lower than 45°F. An engineer is asked to determine whether this requirement will make necessary the installation of an electric preheat coil. If the winter outside design temperature is 5°F and the classroom return air is 70°F, auxiliary electric heat

(A) is unnecessary

(B) should be at least 1.6 kW

(C) should be at least 7.0 kW

(D) should be at least 13 kW

Hint: The default ventilation rate for school classrooms (age 5–8) is the combined total of 10 ft^3/min per person plus 0.12 ft^3/min/ft^2, and the default occupant density is 25 persons per 1000 ft^2.

PROBLEM 57

The heat pump selected for a project has a nominal heating capacity of 60,000 Btu/hr at 47°F outdoor dry-bulb temperature. The manufacturer has provided the integrated heating capacity ratings shown.

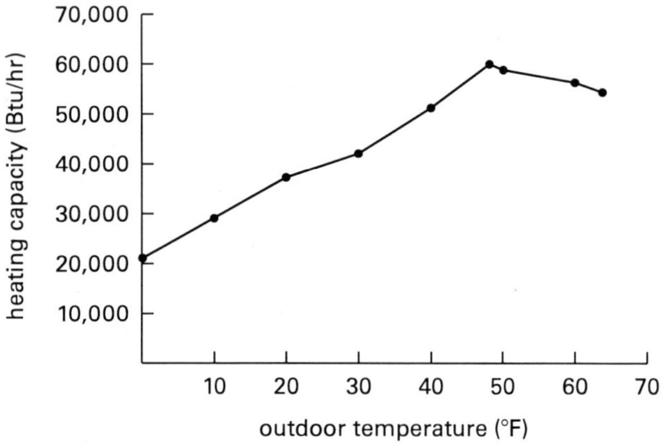

The facility to be heated requires 55,000 Btu/hr at outdoor design conditions of 10°F. For this particular application, auxiliary backup electric heat

(A) is unnecessary

(B) should be at least 7.6 kW

(C) should be at least 16 kW

(D) should be at least 26 kW

Hint: The supplemental heat required is that which is *not* provided by the heat pump at 10°F outdoor conditions.

PROBLEM 58

The ventilation rate procedure given in ANSI/ASHRAE Standard 62.1, *Ventilation for Acceptable Indoor Air Quality*,

(A) requires monitoring of sulfur dioxide, carbon monoxide, and ozone

(B) requires air treatment to control offending contaminants

(C) specifies acceptable concentrations of indoor air contaminants

(D) prescribes the ventilation rate of outdoor air to a space

Hint: There are two procedures that may be used to show compliance with the ventilation standard. The ventilation rate procedure requirements are different than those of the indoor air quality procedure.

PROBLEM 59

When speaking of heat recovery, the parameter ϵ stands for effectiveness. This means

(A) the maximum possible transfer of moisture or energy compared to the Carnot cycle

(B) the actual transfer of moisture or energy compared to the Carnot cycle

(C) the ratio of sensible heat transfer to total heat transfer (sensible and latent)

(D) the actual transfer of moisture or energy compared to the maximum possible

Hint: As with most performance or efficiency parameters, this compares what is actually achieved to what is possible.

PROBLEM 60

A piece of mechanical equipment rotates at 700 rpm and weighs 2500 lbm. The equipment is installed on isolators with stiffness of 2000 lbf/in, causing 1.25 in of deflection. The equipment develops a force of 150 lbf. The force transmitted to the structure is most nearly

(A) 0.0024 lbf

(B) 7.2 lbf

(C) 9.2 lbf

(D) 47 lbf

Hint: The transmissibility is inversely proportional to the square of the ratio of the disturbing frequency to the system's natural frequency.

PROBLEM 61

The preferred method for evaluating HVAC system-related sound is the

(A) room criteria method

(B) A-weighted sound level method

(C) noise criteria method

(D) balanced noise criteria method

Hint: The answer can be found in the ASHRAE Fundamentals Handbook.

PROBLEM 62

A photocopy machine is located 12 ft from the reception desk in a normally furnished 15 ft by 15 ft office space with a 9 ft ceiling. The machine emits 100 dB of sound. For the 1000 Hz octave band, the sound pressure level experienced at the reception desk is most nearly

(A) 41 dB

(B) 64 dB

(C) 89 dB

(D) 120 dB

Hint: The sound power level reduces from the source logarithmically as a function of room volume, octave band center frequency, and distance from the source.

PROBLEM 63

The cost of repairing a recently disabled electric resistance heating system is estimated to be \$40,000. This system uses an average of 640,000 kW-hr per year for heating. As an alternative to the repair, the contractor has proposed the installation of a natural gas-fired system for \$185,000. The gas system has an annual fuel utilization efficiency of 76%. The current cost of electricity is \$0.08/kW-hr and natural gas is \$0.75 per therm. The simple payback for installing a new system rather than repairing the old one is most nearly

(A) 3.7 yr

(B) 4.9 yr

(C) 6.2 yr

(D) 6.7 yr

Hint: Simple payback of an investment is the additional cost divided by the savings per year.

PROBLEM 64

The management of a factory is approached by the vendor of a boiler company with a proposal to install a new boiler. The annual fuel utilization efficiency (AFUE) of the proposed boiler is 0.84, compared to 0.75 for the present boiler, which at present consumes 200,000 therms per year in natural gas. The new boiler will cost \$85,000, and if purchased, the money to pay for it will be borrowed at 8% interest. If the cost of natural gas remains relatively constant at 65 cents per therm, the length of time it will take to recover the investment is most nearly

(A) 6.1 yr

(B) 8.7 yr

(C) 9.3 yr

(D) 11 yr

Hint: The number of periods, n, needed to recover a present investment, P, with an annual recovery, A, at an interest rate, i, is $n = -\log(1 - Pi/A)/\log(1 + i)$.

PROBLEM 65

A landlord plans to borrow a sum of money at 6% annual interest (compounded monthly) to make physical plant improvements to his facility. To determine the monthly payment that will pay off the loan in 5 yr, the correct recovery calculation is

(A) $\dfrac{A}{F} = \dfrac{(0.06)(1+0.06)^5}{(1+0.06)^5 - 1}$

(B) $\dfrac{A}{P} = \dfrac{(0.06)(1+0.06)^5}{(1+0.06)^5 - 1}$

(C) $\dfrac{A}{P} = \dfrac{(0.005)(1+0.005)^{60}}{(1+0.005)^{60} - 1}$

(D) $\dfrac{P}{A} = \dfrac{(1+0.005)^{60} - 1}{(0.005)(1+0.005)^{60}}$

Hint: The landlord wants a loan whose monthly payment is A for present value improvements of P.

SOLUTION 1

Use the ASHRAE sea level psychrometric chart. Plot the psychrometric state point of the entering room air, and then plot the path of the room sensible/latent process line to the final dry-bulb temperature of the room. That interception point indicates the final state point, which includes the humidity ratio.

The slope of the room sensible/latent process line is called the sensible heat ratio and is calculated as

$$\begin{aligned}\text{SHR} &= \frac{Q_{\text{sensible}}}{Q_{\text{total}}} = \frac{Q_{\text{sensible}}}{Q_{\text{sensible}} + Q_{\text{latent}}} \\ &= \frac{100{,}000\ \frac{\text{Btu}}{\text{hr}}}{100{,}000\ \frac{\text{Btu}}{\text{hr}} + 45{,}000\ \frac{\text{Btu}}{\text{hr}}} \\ &= 0.69\end{aligned}$$

Most psychrometric charts have a "protractor" or other means of plotting a line with the required sensible heat ratio slope. Once established, this slope can be graphically transferred to the supply air state point and extended so that it intercepts the 75°F temperature line. The process is shown here with only relevant lines given.

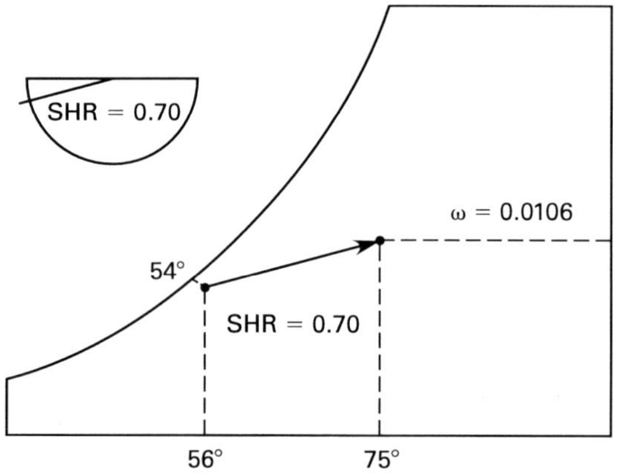

As shown, the point of interception has a humidity ratio of

$$\omega = 0.0102\ \text{lbm/lbm} \quad (0.010\ \text{lbm/lbm})$$

The answer is (C).

Why Other Options Are Wrong

(A) This answer results from a common mistake made when working with the very small units of lbm/lbm. The answer has been misread from the chart by one decimal place.

(B) This incorrect answer results from ignoring the latent component of the room load and heating sensibly to 75°F.

(D) This incorrect answer results from plotting the ratio of the latent to the total load.

SOLUTION 2

This problem is most readily solved on the psychrometric chart. The sensible heat ratio is

$$\text{SHR} = \frac{Q_{\text{sensible}}}{Q_{\text{sensible}} + Q_{\text{latent}}}$$

The sensible heat gain is

$$\begin{aligned}Q_{\text{sensible}} &= \text{lighting gain} \\ &\quad + \text{sensible heat gain from people} \\ &= (10\ \text{kW})\left(\frac{3412\ \frac{\text{Btu}}{\text{hr}}}{1\ \text{kW}}\right) \\ &\quad + (300\ \text{persons})\left(\frac{200\ \frac{\text{Btu}}{\text{hr}}}{1\ \text{person}}\right) \\ &= 94{,}120\ \text{Btu/hr}\end{aligned}$$

The latent heat gain is

$$\begin{aligned}Q_{\text{latent}} &= (300\ \text{persons})\left(\frac{150\ \frac{\text{Btu}}{\text{hr}}}{1\ \text{person}}\right) \\ &= 45{,}000\ \text{Btu/hr}\end{aligned}$$

Thus, the sensible heat ratio is

$$\begin{aligned}\text{SHR} &= \frac{94{,}120\ \frac{\text{Btu}}{\text{hr}}}{94{,}120\ \frac{\text{Btu}}{\text{hr}} + 45{,}000\ \frac{\text{Btu}}{\text{hr}}} \\ &= 0.6765 \quad (0.68)\end{aligned}$$

These results are plotted on the psychrometric chart as shown.

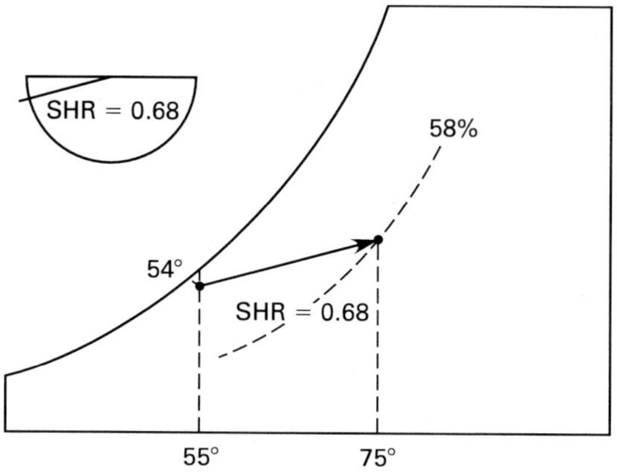

Using an ASHRAE chart, the SHR is first plotted on the protractor provided to determine the graphical slope of the process. Start at the entering conditions of 55°F dry-bulb and 54°F wet-bulb temperature, and draw a line with a slope of approximately 0.68 (0.7 for practical purposes) to the vertical 75°F dry-bulb line. At that termination, the relative humidity is approximately 58% (rounded to 60%).

The answer is (B).

Why Other Options Are Wrong

(A) This incorrect answer might be selected if the sensible heat ratio is ignored and a horizontal line is plotted to a 75°F termination.

(C) This answer results from calculating the slope incorrectly, taking the ratio of the latent load to the total load.

(D) This incorrect answer results from calculating the state point but not proceeding along the process line.

SOLUTION 3

The reheat load is represented by the increase in enthalpy that must occur between the entering and leaving conditions of the air.

$$\begin{aligned} Q_{\text{reheat}} &= \dot{m}(h_{\text{out}} - h_{\text{in}}) \\ &= \rho \dot{V}(h_{\text{out}} - h_{\text{in}}) \end{aligned}$$

First locate the state point of the air leaving the coil (as in the following psychrometric chart). Follow the process 2°F to the right to account for the fan heat imposed on the airstream and to locate the state point of the air entering the reheat coil. From the chart, the enthalpy of the air leaving the fan (entering the coil) is

$$h_{\text{in}} = 17.1 \text{ Btu/lbm}$$

Also from the chart, the density of the same air is

$$\begin{aligned} \rho &= \frac{1}{v} \\ &= \frac{1}{12.8 \ \frac{\text{ft}^3}{\text{lbm}}} \\ &= 0.078 \text{ lbm/ft}^3 \end{aligned}$$

The reheat process is a sensible one, so the process line continues horizontally to the right until it intercepts the desired temperature condition of 55°F. The enthalpy of the air leaving the reheat coil is

$$h_{\text{out}} = 19.3 \text{ Btu/lbm}$$

The psychrometric process can be depicted schematically as

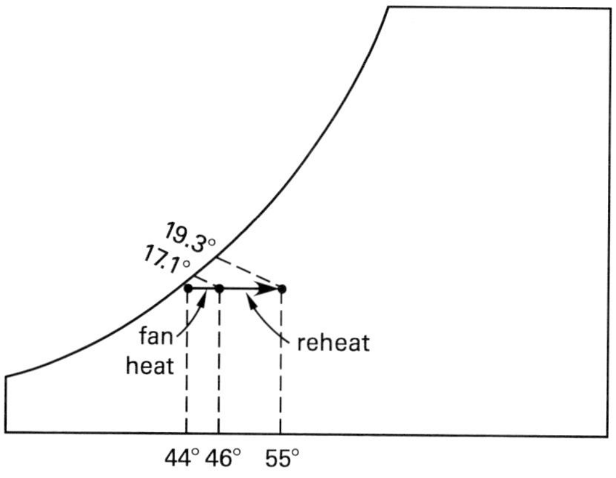

The reheat load is

$$\begin{aligned} Q_{\text{reheat}} &= \rho \dot{V}(h_{\text{out}} - h_{\text{in}}) \\ &= \left(0.078 \ \frac{\text{lbm}}{\text{ft}^3}\right)\left(2000 \ \frac{\text{ft}^3}{\text{min}}\right)\left(60 \ \frac{\text{min}}{\text{hr}}\right) \\ &\quad \times \left(19.3 \ \frac{\text{Btu}}{\text{lbm}} - 17.1 \ \frac{\text{Btu}}{\text{lbm}}\right) \\ &= 20{,}592 \text{ Btu/hr} \quad (20{,}600 \text{ Btu/hr}) \end{aligned}$$

The answer is (B).

Why Other Options Are Wrong

(A) This incorrect answer results from using the density of air at standard conditions instead of the conditions indicated in the process.

(C) This incorrect answer results from failing to account for the temperature rise across the fan, which reduces the reheat load.

(D) This incorrect answer results from reheating the air to room conditions, rather than to the 55°F specified for the supply air.

Depth

SOLUTION 4

The total heat that must be removed to lower the temperature of the pork shoulder from 40°F to 10°F is the sum of the sensible heat that must be removed above and below freezing, as well as the latent heat which must be removed to freeze the product.

$$q_{\text{total}} = q_{\text{sensible,above}} + q_{\text{latent}} + q_{\text{sensible,below}}$$

The sensible heat required to lower the temperature of the product from its original temperature to its freezing point is

$$\begin{aligned} q_{\text{sensible,above}} &= m_{\text{meat}} c_p (T_0 - T_1) \\ &= (10{,}000 \text{ lbm})\left(0.86 \, \frac{\text{Btu}}{\text{lbm-°F}}\right) \\ &\quad \times (40°\text{F} - 28°\text{F}) \\ &= 103{,}200 \text{ Btu} \end{aligned}$$

The latent heat required to freeze the product is

$$q_{\text{latent}} = m_{\text{water}} h_f$$

From a table of water properties, the latent heat of fusion for water at atmospheric pressure is

$$h_f = 144 \text{ Btu/lbm}$$

The mass of the water is obtained by multiplying the mass of the meat by its percentage of water content. The latent heat of fusion for the product is thus

$$\begin{aligned} q_{\text{latent}} &= m_{\text{meat}} w_{\%} h_f = (10{,}000 \text{ lbm})(0.73)\left(144 \, \frac{\text{Btu}}{\text{lbm}}\right) \\ &= 1{,}051{,}200 \text{ Btu} \end{aligned}$$

The sensible heat required to lower the temperature of the product from its freezing point to its final temperature is

$$\begin{aligned} q_{\text{sensible,below}} &= m_{\text{meat}} c_p (T_1 - T_2) \\ &= (10{,}000 \text{ lbm})\left(0.53 \, \frac{\text{Btu}}{\text{lbm-°F}}\right) \\ &\quad \times (28°\text{F} - 10°\text{F}) \\ &= 95{,}400 \text{ Btu} \end{aligned}$$

The cooling capacity must be enough to remove the total heat within 2 hr.

$$\begin{aligned} Q &= \frac{q_{\text{total}}}{t} \\ &= \frac{q_{\text{sensible,above}} + q_{\text{latent}} + q_{\text{sensible,below}}}{t} \\ &= \frac{103{,}200 \text{ Btu} + 1{,}051{,}200 \text{ Btu} + 95{,}400 \text{ Btu}}{2 \text{ hr}} \\ &= 624{,}900 \text{ Btu/hr} \quad (625{,}000 \text{ Btu/hr}) \end{aligned}$$

The answer is (A).

Why Other Options Are Wrong

(B) This answer is wrong because the specific heats for above and below freezing have been reversed in the application.

(C) This answer is wrong because the percentage of water content was left out of the calculation.

(D) This answer is wrong because the total heat was not divided by 2 hr.

SOLUTION 5

The energy requirement for any psychrometric process is calculated as

$$\begin{aligned} Q &= \dot{m}\Delta h \\ &= \rho \dot{V} \Delta h \\ &= \left(\frac{\dot{V}}{v}\right)\Delta h \end{aligned}$$

It is not possible to go directly to the desired psychrometric state point with a conventional cooling coil, so it is necessary to first cool and dehumidify the air to the same humidity ratio as the final state point, and then reheat the air sensibly to the desired final dry-bulb temperature. The energy required for the total process is

$$Q_{\text{process}} = Q_{\text{cool}} + Q_{\text{reheat}}$$

The energy needed by the refrigeration system must be adjusted by the energy efficiency rating. The efficiencies of the coils are nearly 100% and may be disregarded.

The energy needs for the individual processes are

$$\begin{aligned} Q_{\text{cool}} &= \frac{\dot{V}}{v}(h_1 - h_2)\left(\frac{1}{\text{EER}}\right) \\ Q_{\text{reheat}} &= \frac{\dot{V}}{v}(h_3 - h_2) \end{aligned}$$

From the psychrometric chart, at the initial state point of 75°F dry-bulb/60% relative humidity, the specific volume is 13.7 ft^3/lbm and the enthalpy is

$$h_1 = 30.2 \text{ Btu/lbm of dry air}$$

The final state point of 75°F dry-bulb/50% relative humidity has a humidity ratio of 0.0093 lbm/lbm and an enthalpy of

$$h_3 = 28.1 \text{ Btu/lbm of dry air}$$

If the coil efficiency is very nearly 100%, the cooling process will be from 75°F dry-bulb/60% relative humidity to the point on the saturation curve with a humidity ratio of 0.0093 lbm/lbm. This intermediate point has an enthalpy of

$$h_2 = 23.2 \text{ Btu/lbm of dry air}$$

The energy rate for the cooling process is

$$Q_{\text{cool}} = \frac{\left(1500 \ \frac{\text{ft}^3}{\text{min}}\right)\left(60 \ \frac{\text{min}}{\text{hr}}\right)}{13.7 \ \frac{\text{ft}^3}{\text{lbm}}} \times \left(30.2 \ \frac{\text{Btu}}{\text{lbm}} - 23.2 \ \frac{\text{Btu}}{\text{lbm}}\right)\left(\frac{1 \text{ W}}{7.5 \ \frac{\text{Btu}}{\text{hr}}}\right) = 6131 \text{ W}$$

The energy rate for the reheating process is

$$Q_{\text{reheat}} = \frac{\left(1500 \ \frac{\text{ft}^3}{\text{min}}\right)\left(60 \ \frac{\text{min}}{\text{hr}}\right)}{13.7 \ \frac{\text{ft}^3}{\text{min}}} \times \left(28.1 \ \frac{\text{Btu}}{\text{lbm}} - 23.2 \ \frac{\text{Btu}}{\text{lbm}}\right) = 32{,}190 \text{ Btu/hr}$$

Converted to watts, this is

$$Q_{\text{reheat}} = \left(32{,}190 \ \frac{\text{Btu}}{\text{hr}}\right)\left(\frac{1 \text{ W}}{3.412 \ \frac{\text{Btu}}{\text{hr}}}\right) = 9434 \text{ W}$$

The total process electrical requirement is

$$\begin{aligned} Q_{\text{process}} &= 6131 \text{ W} + 9434 \text{ W} \\ &= 15{,}565 \text{ W} \quad (16{,}000 \text{ W}) \end{aligned}$$

The answer is (C).

Why Other Options Are Wrong

(A) This incorrect answer results from considering only the cooling process.

(B) This incorrect answer results from considering only the reheating process.

(D) This incorrect answer results from neglecting to account for the energy efficiency rating in the cooling process.

SOLUTION 6

The minimum humidification rate is given by

$$\begin{aligned} \dot{m}_{\text{water}} &= \dot{m}_{\text{air}}(\omega_{\text{in}} - \omega_{\text{out}}) \\ &= \rho_{\text{air}} \dot{V}_{\text{air}}(\omega_{\text{in}} - \omega_{\text{out}}) \end{aligned}$$

The minimum volume flow rate is

$$\begin{aligned} \dot{V}_{\text{air}} &= \dot{n}_{\text{changes}} \\ &= \left(8.0 \ \frac{\text{air changes}}{\text{hr}}\right)\left(\frac{(20 \text{ ft})(25 \text{ ft})(9 \text{ ft})}{1 \text{ air change}}\right) \\ &= 36{,}000 \text{ ft}^3/\text{hr} \end{aligned}$$

The ASHRAE low temperature psychrometric chart indicates that at 10°F dry-bulb and 8°F wet-bulb, the humidity ratio is

$$\omega_{\text{out}} = 0.00080 \text{ lbm/lbm}$$

The humidity ratio at 70°F dry-bulb and 40% relative humidity is

$$\omega_{\text{in}} = 0.0062 \text{ lbm/lbm}$$

The minimum humidification rate is

$$\begin{aligned} \dot{m}_{\text{water}} &= \rho_{\text{air}} \dot{V}_{\text{water}}(\omega_{\text{in}} - \omega_{\text{out}}) \\ &= \left(0.074 \ \frac{\text{lbm}}{\text{ft}^3}\right)\left(36{,}000 \ \frac{\text{ft}^3}{\text{hr}}\right) \times \left(0.0062 \ \frac{\text{lbm}}{\text{lbm}} - 0.0008 \ \frac{\text{lbm}}{\text{lbm}}\right) \\ &= 14.38 \text{ lbm/hr} \quad (15 \text{ lbm/hr}) \end{aligned}$$

The answer is (B).

Why Other Options Are Wrong

(A) This incorrect answer results from misreading the outdoor humidity ratio as 0.008 and then reversing the two humidity ratios in order to subtract the smaller from the larger.

(C) This incorrect answer results from using the density of the outside air to convert from volume flow to mass flow. The right answer depends on getting the mass flow rate correct. The criterion of eight air changes per hour is defined (and physically verified in hospitals) for the indoor condition. The density of the air used to establish mass flow must be taken at the same indoor conditions used to determine volume flow.

(D) This incorrect answer results from expressing the volume flow in cubic feet per minute (frequently done but unnecessary here) and neglecting to convert to cubic feet per hour for the final answer.

SOLUTION 7

The steam humidification process will follow a process line on the psychrometric chart that has a slope equal to $\Delta h/\Delta\omega$.

From the steam tables, saturated steam at 15 psig (or 29.7 psia) has a specific enthalpy of 1163.9 Btu/lbm. The condition line can be drawn using the protractor provided on the ASHRAE psychrometric chart. After transferring a line of the proper slope from the protractor to the initial condition plotted on the chart as shown, a process line starting at 70°F and 20% relative humidity is run until it intersects the 50°F dew point line.

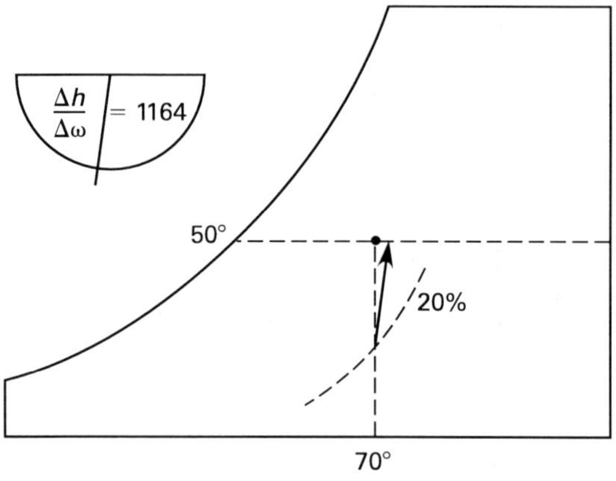

This represents the final state of the humidified air. The dry-bulb temperature is $T_{\text{final}} = 72°\text{F}$.

The answer is (D).

Alternate Method

If the psychrometric chart being used does not have a protractor, the change in enthalpy between the two states can be determined by multiplying the slope (1163.9 Btu/lbm, as found in the previous solution) by the change in humidity ratio ($\Delta\omega$). The change in humidity ratio is the difference between the humidity ratio coincident with the 50°F dew-point line and the humidity ratio air at of 70°F and 20% relative humidity.

$$\begin{aligned}\Delta h &= \Delta\omega\left(\frac{\Delta h}{\Delta\omega}\right) = (\omega_1 - \omega_2)(\text{slope})\\ &= \left(0.00766\ \frac{\text{lbm}}{\text{lbm}} - 0.00310\ \frac{\text{lbm}}{\text{lbm}}\right)\left(1163.9\ \frac{\text{Btu}}{\text{lbm}}\right)\\ &= 5.307\ \text{Btu/lbm}\end{aligned}$$

Adding this enthalpy increase to the initial enthalpy (20.1885 Btu/lbm) yields the final enthalpy of the moist air.

$$\begin{aligned}h_{\text{final}} &= h_{\text{initial}} + \Delta h\\ &= 20.1885\ \frac{\text{Btu}}{\text{lbm}} + 5.307\ \frac{\text{Btu}}{\text{lbm}}\\ &= 25.5\ \text{Btu/lbm}\end{aligned}$$

The intersection of the final enthalpy with the 50°F dew-point line represents the final state point of the moist air.

$$T_{\text{final}} = 72°\text{F}$$

The answer is (D).

Why Other Options Are Wrong

(A) This answer results from incorrectly assuming that the final state of the air is saturated at a 50°F dew point and drawing the process line from the initial conditions directly to the final dew point.

(B) This incorrect answer results from using the common approximation of 1000 Btu/lbm for the energy of latent heat. In this case, the error in slope on the psychrometric chart is considerable.

(C) This incorrect answer results from confusing change in enthalpy with pure change in humidity ratio, a latent process. The line is simply drawn vertically to its intersection with the 50°F dew-point line.

SOLUTION 8

For swimming pool environments, the 1995 ASHRAE HVAC Applications Handbook recommends that makeup air requirements be calculated with the equation

$$\dot{V} = \frac{w_p}{\rho(\omega_i - \omega_o)}$$

(This equation does not appear in later editions, but the approach is still standard.) From the psychrometric chart, the humidity ratio for the outside air at 90°F dry-bulb and 68°F wet-bulb temperature is 0.0097 lbm/lbm. The humidity ratio for the indoor pool room air at 82°F and 60% relative humidity is 0.014 lbm/lbm.

The required outside air is

$$\dot{V} = \frac{350\ \frac{\text{lbm}}{\text{hr}}}{\left(0.075\ \frac{\text{lbm}}{\text{ft}^3}\right)\left(0.014\ \frac{\text{lbm}}{\text{lbm}} - 0.0097\ \frac{\text{lbm}}{\text{lbm}}\right)} \times \left(60\ \frac{\text{min}}{\text{hr}}\right)$$
$$= 18{,}087\ \text{ft}^3/\text{min} \quad (18{,}100\ \text{ft}^3/\text{min})$$

The answer is (C).

Why Other Options Are Wrong

(A) This incorrect answer results from neglecting to include the density term.

(B) This incorrect answer results from using the industry shortcut $q_{\text{latent}} = 4840\,\dot{V}\Delta\omega$.

This is an unnecessary rule of thumb that introduces errors. The shortcut is based on an enthalpy change from 68°F (1055.5 Btu/lbm), rather than a value for the 80°F pool.

(D) This incorrect answer results from neglecting the conversion from minutes to hours.

SOLUTION 9

For systems with variable outside air quantities, the equation to calculate the humidification load specifically recommended by the ASHRAE HVAC Systems and Equipment Handbook is

$$H = 60\rho\dot{V}(\omega_i - \omega_o)\left(\frac{T_i - T_m}{T_i - T_o}\right) - S + L$$

For this equation, ASHRAE defines ρ as the density of air at sea level, 0.074 lbm/ft^3. S and L represent other moisture additions and losses that are not present in this problem. T_m is the mixed air temperature, which can be found using the common psychrometric practice of "weighting" the properties of the two airstreams.

$$T_m = (75°\text{F})\left(\frac{6000\ \frac{\text{ft}^3}{\text{min}}}{10{,}000\ \frac{\text{ft}^3}{\text{min}}}\right) + (10°\text{F})\left(\frac{4000\ \frac{\text{ft}^3}{\text{min}}}{10{,}000\ \frac{\text{ft}^3}{\text{min}}}\right)$$
$$= 49°\text{F}$$

From the psychrometric chart, indoor conditions of 75°F/40% relative humidity give an indoor humidity ratio (ω_i) of 0.0074 lbm/lbm. Outdoor conditions of 10°F dry-bulb /8°F wet-bulb give an outdoor humidity ratio (ω_o) of 0.0008 lbm/lbm.

For the problem at hand, the humidification load is

$$H = \rho\dot{V}(\omega_i - \omega_o)\left(\frac{T_i - T_m}{T_i - T_o}\right)$$
$$= \left(0.074\ \frac{\text{lbm}}{\text{ft}^3}\right)\left(10{,}000\ \frac{\text{ft}^3}{\text{min}}\right)\left(60\ \frac{\text{min}}{\text{hr}}\right) \times \left(0.0074\ \frac{\text{lbm}}{\text{lbm}} - 0.0008\ \frac{\text{lbm}}{\text{lbm}}\right)\left(\frac{75°\text{F} - 49°\text{F}}{75°\text{F} - 10°\text{F}}\right)$$
$$= 117\ \text{lbm/hr} \quad (120\ \text{lbm/hr})$$

The answer is (D).

Why Other Options Are Wrong

(A) This incorrect answer results from neglecting to convert minutes to hours.

(B) This incorrect answer results from misreading the decimal place for the outside air humidity ratio and then transposing the two terms in the difference calculation.

(C) This incorrect answer results from using the outside air quantity, even though the equation is given in a form to be used with the mixed air *total* results. In other words, the fraction of outside air as well as preheating has already been accounted for in the mixed air temperature given.

SOLUTION 10

Adopting the convention that the water surface in the tank is point 1 and the water in the pipe at the pressure gauge is point 2, write the Bernoulli equation as

$$\frac{p_1 g_c}{\rho g} + \frac{\text{v}_1^2}{2g} + z_1 + h_M = \frac{p_2 g_c}{\rho g} + \frac{\text{v}_2^2}{2g} + z_2$$

At point 1 the velocity is 0 ft/sec, the elevation is 9 ft, and the static head is atmospheric (0 psig). At point 2 the static head is 23 psig and the elevation head is 0 ft. The velocity at point 2 can now be described as

$$\mathrm{v}_2 = \sqrt{\left(\frac{(p_1 - p_2)g_c}{\rho g} + z_1 + h_M\right)2g}$$

$$= \sqrt{\begin{array}{l}\left(\dfrac{\left(0\ \dfrac{\text{lbf}}{\text{in}^2} - 23\ \dfrac{\text{lbf}}{\text{in}^2}\right)\left(12\ \dfrac{\text{in}}{\text{ft}}\right)^2\left(32.2\ \dfrac{\text{ft-lbm}}{\text{lbf-sec}^2}\right)}{\left(62.34\ \dfrac{\text{lbm}}{\text{ft}^3}\right)\left(32.2\ \dfrac{\text{ft}}{\text{sec}^2}\right)} \right.\\ \left. \quad +9\ \text{ft} + 46\ \text{ft}\right) \\ \quad \times(2)\left(32.2\ \dfrac{\text{ft}}{\text{sec}^2}\right)\end{array}}$$

$$= 11\ \text{ft/sec}$$

The answer is (B).

Why Other Options Are Wrong

(A) This incorrect answer results when the $2g$ is dropped from the velocity head term.

(C) This incorrect answer results when the conversion factor g_c is neglected.

(D) This incorrect answer results when the final step of taking the square root is omitted.

SOLUTION 11

The Darcy-Weisbach equation is stated as

$$\Delta p = f\left(\frac{L}{d}\right)\left(\frac{g}{g_c}\right)\left(\frac{\mathrm{v}^2}{2}\right)$$

This relationship shows that the fluid friction is directly affected by the Moody friction factor, f, the length of the conduit, L, and the velocity, v. It is inversely proportional to the diameter of the conduit, d.

The answer is (B).

Why Other Options Are Wrong

(A) The first law of thermodynamics gives rise to the Bernoulli equation but provides no basis for calculating pressure drop. This answer is incorrect.

(C) While the Bernoulli equation is considered to be the basic tool of flow analysis, it merely describes how to apply losses due to friction. It does not show how to calculate them.

(D) Computational fluid dynamics is a powerful tool for analyzing fluid motion. But it is not used directly to determine friction loss in a conduit.

SOLUTION 12

The pressure gauges can only read static head. Assuming friction and fitting losses to be negligible for the short distance, the Bernoulli equation in terms of pressure can be stated as

$$p_1 + \frac{\rho \mathrm{v}_1^2}{2g_c} + \frac{\rho g z_1}{g_c} + h_{\mathrm{M}} = p_2 + \frac{\rho \mathrm{v}_2^2}{2g_c} + \frac{\rho g z_2}{g_c}$$

Rearranging to solve for the static discharge head of the pump results in

$$p_2 = p_1 + \left(\frac{\rho \mathrm{v}_1^2}{2g_c} - \frac{\rho \mathrm{v}_2^2}{2g_c}\right) + \left(\frac{\rho g z_1}{g_c} - \frac{\rho g z_2}{g_c}\right) + h_{\mathrm{M}}$$

The velocity terms are equal (if the pipe size remains constant), and the elevation terms are equal, simplifying the equation to

$$p_2 = p_1 + h_{\mathrm{M}}$$

The pressure on the discharge gauge is the sum of the pump inlet head and the total dynamic head of the pump. In units of psig, this is

$$p_2 = 5\ \frac{\text{lbf}}{\text{in}^2} + (60\ \text{ft wg})\left(\frac{0.43352\ \dfrac{\text{lbf}}{\text{in}^2}}{1\ \text{ft wg}}\right)$$

$$= 31\ \text{psig}$$

The answer is (A).

Why Other Options Are Wrong

(B) This incorrect answer results from including the final velocity head term in the solution. This head is not detected by the pressure gauge, which only senses static pressure.

(C) This incorrect answer results from converting the final units to psia instead of psig.

(D) This incorrect answer results from omitting the unit conversion from dynamic head to psig.

SOLUTION 13

The heat loss for a surface exposed to indoor and outdoor ambient conditions is the product of the U-factor, the surface area, and the temperature difference across the entire component.

$$Q = UA(T_{\text{in}} - T_{\text{out}})$$

Three surfaces in the room face ambient outdoor conditions. The room heat loss is the sum of the heat losses of these three surfaces.

$$\begin{aligned} Q_{\text{room}} &= \sum U_i A_i (T_{\text{in}} - T_{\text{out}}) \\ &= \left(0.094 \ \frac{\text{Btu}}{\text{hr-ft}^2\text{-}^\circ\text{F}}\right)(120 \ \text{ft}^2)(70^\circ\text{F} - 15^\circ\text{F}) \\ &\quad + \left(0.65 \ \frac{\text{Btu}}{\text{hr-ft}^2\text{-}^\circ\text{F}}\right)(24 \ \text{ft}^2)(70^\circ\text{F} - 15^\circ\text{F}) \\ &\quad + \left(0.037 \ \frac{\text{Btu}}{\text{hr-ft}^2\text{-}^\circ\text{F}}\right)(200 \ \text{ft}^2)(70^\circ\text{F} - 15^\circ\text{F}) \\ &= 1885.4 \ \text{Btu/hr} \end{aligned}$$

Converting to watts,

$$\begin{aligned} Q_{\text{room}} &= \left(1885.4 \ \frac{\text{Btu}}{\text{hr}}\right)\left(\frac{1 \ \text{W-hr}}{3.412 \ \text{Btu}}\right) \\ &= 552.58 \ \text{W} \end{aligned}$$

The length of electric baseboard heating required is the room heat loss divided by the unit (per-foot) capacity of the baseboard heater.

$$\begin{aligned} L &= \frac{Q_{\text{room}}}{C_{\text{per ft}}} \\ &= \frac{552.58 \ \text{W}}{250 \ \frac{\text{W}}{\text{ft}}} \\ &= 2.21 \ \text{ft} \quad (3 \ \text{ft}) \end{aligned}$$

The answer is (B).

Why Other Options Are Wrong

(A) While close to the calculated requirement, a 2 ft length would be insufficient to maintain a temperature of 70°F when the outdoor temperature is 15°F.

(C) This incorrect answer results from a slip of a decimal place, taking the U-factor for the ceiling as 0.37.

(D) This incorrect answer results from failing to convert Btu/hr to watts.

SOLUTION 14

The glazing solar cooling load is

$$\begin{aligned} Q_{\text{glazing}} &= A(\text{SCL})(\text{SC}) \\ &= (1500 \ \text{ft}^2)\left(193 \ \frac{\text{Btu}}{\text{hr-ft}^2}\right)(0.81) \\ &= 234{,}495 \ \text{Btu/hr} \end{aligned}$$

The supply airflow is determined from the sensible load and the desired temperature rise between the time the air enters and leaves the room. The required airflow is

$$\begin{aligned} \dot{V} &= \frac{Q_{\text{sensible}}}{\rho c_p (T_{\text{room}} - T_{\text{supply}})} \\ &= \frac{\left(234{,}495 \ \frac{\text{Btu}}{\text{hr}}\right)\left(\frac{1 \ \text{hr}}{60 \ \text{min}}\right)}{\left(0.075 \ \frac{\text{lbm}}{\text{ft}^3}\right)\left(0.24 \ \frac{\text{Btu}}{\text{lbm-}^\circ\text{F}}\right)(74^\circ\text{F} - 58^\circ\text{F})} \\ &= 13{,}570 \ \text{ft}^3/\text{min} \quad (13{,}600 \ \text{ft}^3/\text{min}) \end{aligned}$$

The answer is (A).

Why Other Options Are Wrong

(B) This incorrect answer results from including the latent load in the calculation. Supply air quantities are determined from the sensible load based on dry-bulb temperature, because that is what the thermostat responds to.

(C) This incorrect answer results from leaving the shading coefficient out of the answer.

(D) This incorrect answer results from omitting the conversion from minutes to hours.

SOLUTION 15

The mean radiant temperature is the summation of angle-factor/surface temperature products for the entire "view" the occupant is exposed to.

$$T_r = F_1 T_1 + F_2 T_2 + \ldots + F_n T_n$$

The problem is most easily solved if a shape factor is determined for the cold window surface, F_w, and then all other surfaces in the room represent the balance, F_o. According to ASHRAE Standard 55, the three parameters of importance here are width of the surface (a), height of the surface (b), and distance from the occupant to the surface (c). Taking a as 22 ft, b as 6.5 ft, and c as 5 ft, the following ratios can be calculated.

$$\begin{aligned} \frac{a}{c} &= \frac{22 \ \text{ft}}{5 \ \text{ft}} \\ &= 4.4 \\ \frac{b}{c} &= \frac{6.5 \ \text{ft}}{5 \ \text{ft}} \\ &= 1.3 \end{aligned}$$

Depth

Entering the chart with these ratios indicates that the cold window surface accounts for about 7.5% ($F_w = 0.075$) of the radiation view of the office occupant. The factor for the rest of the room is

$$\begin{aligned} F_o &= 1 - F_w \\ &= 0.925 \end{aligned}$$

The mean radiant temperature experienced by the occupant is the weighted surface temperature of the glass at 40°F and all other surfaces at 70°F.

$$\begin{aligned} T_r &= F_w T_w + F_o T_o \\ &= (0.075)(40°\text{F}) + (1 - 0.075)(70°\text{F}) \\ &= 67.8°\text{F} \quad (68°\text{F}) \end{aligned}$$

The answer is (C).

Why Other Options Are Wrong

(A) This incorrect answer results from assuming that the mean radiant temperature is dominated by the temperature of the nearby window.

(B) This incorrect answer is the simple average of the two temperatures. This would be correct only if the shape factor were coincidentally equal to 0.50.

(D) This incorrect answer results from assuming that the mean radiant temperature is dominated by the air temperature.

SOLUTION 16

ASHRAE recommends that pool evaporation, w_p (in lbm/hr), be estimated using the following empirical equation.

$$w_p = \left(\frac{A}{Y}\right)(p_w - p_a)(95 + 0.425\text{v})$$

The area of the pool is

$$\begin{aligned} A &= Lw = (165 \text{ ft})(45 \text{ ft}) \\ &= 7425 \text{ ft}^2 \end{aligned}$$

The latent heat required to change water to vapor at the pool surface water temperature of 80°F is

$$Y = 1048.4 \text{ Btu/lbm}$$

At 82°F and 55% relative humidity, the room air dew point is approximately 64°F. The saturation pressure at 64°F can be found from a table of thermodynamic properties of moist air, and is

$$p_a = 0.601 \text{ in Hg}$$

The saturation vapor pressure taken at the surface water temperature of 80°F is

$$p_w = 1.033 \text{ in Hg}$$

From the first equation, the estimated evaporation load of the pool is

$$\begin{aligned} w_p &= \left(\frac{7425 \text{ ft}^2}{1048.4 \; \frac{\text{Btu}}{\text{lbm}}}\right)(1.033 \text{ in Hg} - 0.601 \text{ in Hg}) \\ &\quad \times \left(95 + (0.425)\left(25 \; \frac{\text{ft}}{\text{min}}\right)\right) \\ &= 323 \text{ lbm/hr} \quad (320 \text{ lbm/hr}) \end{aligned}$$

The answer is (C).

Why Other Options Are Wrong

(A) This incorrect answer results from taking the saturation pressure at the room's air dry-bulb temperature of 82°F rather than at its dew point of 64°F, and then reversing the order of terms when calculating the pressure difference.

(B) This incorrect answer results from using pressure units of psia (directly from steam tables), rather than converting to in Hg.

(D) This incorrect answer results from reading the pool water latent heat from the pressure section of steam tables (80 psia) rather than from the temperature section (80°F).

SOLUTION 17

The saturation efficiency for an evaporative cooler describes the extent to which the dry-bulb temperature is reduced to the theoretical minimum wet-bulb temperature.

$$\begin{aligned} \eta_{\text{sat}} &= \frac{T_{\text{db,air,in}} - T_{\text{db,air,out}}}{T_{\text{db,air,in}} - T_{\text{wb}}} \\ &= \frac{96°\text{F} - 62°\text{F}}{96°\text{F} - 55°\text{F}} \\ &= 0.83 \quad (83\%) \end{aligned}$$

The answer is (C).

Why Other Options Are Wrong

(A) This incorrect answer results from using the temperature differential between the water and the leaving air temperature in the numerator.

(B) This incorrect answer results from using the water-spray temperature rather than the outdoor wet-bulb temperature.

(D) This incorrect answer results from reasoning that a spray water temperature lower than the wet-bulb would result in a lower leaving temperature than wet-bulb, and therefore an efficiency greater than 100% for the definition given.

SOLUTION 18

The relationship between fluid stream heat transfer, mass flow rate, and temperature is

$$Q = \dot{m} c_p \Delta T$$

The temperature drop of the fluid between the heater and the load is

$$\begin{aligned}\Delta T_1 &= T_{\text{heater}} - T_{\text{load}} = \frac{Q}{\dot{m} c_p} \\ &= \frac{Q}{\rho \dot{V} c_p} \\ &= \frac{100{,}000\ \frac{\text{Btu}}{\text{hr}}}{\left(62.34\ \frac{\text{lbm}}{\text{ft}^3}\right)\left(104\ \frac{\text{gal}}{\text{min}}\right) \times \left(\frac{1\ \text{ft}^3}{7.5\ \text{gal}}\right)\left(60\ \frac{\text{min}}{\text{hr}}\right)\left(1.0\ \frac{\text{Btu}}{\text{lbm-°F}}\right)} \\ &= 1.93\text{°F}\end{aligned}$$

The temperature of the water as it is delivered (after piping losses) is therefore

$$\begin{aligned}T_{\text{load}} &= T_{\text{heater}} - \Delta T_1 \\ &= 125\text{°F} - 1.93\text{°F} \\ &= 123.07\text{°F}\end{aligned}$$

After the demand of 44 gpm is satisfied, the remaining 60 gpm of 123.07°F water is returned to the mechanical room, after experiencing additional piping losses of 82,000 Btu/hr. The temperature drop during the return to the heater is

$$\begin{aligned}\Delta T_2 &= \frac{Q}{\rho \dot{V} c_p} \\ &= \frac{82{,}000\ \frac{\text{Btu}}{\text{hr}}}{\left(62.34\ \frac{\text{lbm}}{\text{ft}^3}\right)\left(60\ \frac{\text{gal}}{\text{min}}\right)\left(\frac{1\ \text{ft}^3}{7.5\ \text{gal}}\right) \times \left(60\ \frac{\text{min}}{\text{hr}}\right)\left(1.0\ \frac{\text{Btu}}{\text{lbm-°F}}\right)} \\ &= 2.74\text{°F}\end{aligned}$$

The return water temperature is

$$\begin{aligned}T_{\text{return}} &= T_{\text{load}} - \Delta T_2 \\ &= 123.07\text{°F} - 2.74\text{°F} \\ &= 120.33\text{°F}\end{aligned}$$

The temperature of the water mixture flowing to the heater is the mass-weighted temperature of the return water and makeup water. For constant density and specific heat, no loss in accuracy is experienced by using flow-weighted temperatures.

$$\begin{aligned}T_{\text{heater, in}} &= \frac{\dot{m}_{\text{return}} T_{\text{return}} + \dot{m}_{\text{makeup}} T_{\text{makeup}}}{\dot{m}_{\text{return}} + \dot{m}_{\text{makeup}}} \\ &\approx \frac{\dot{V}_{\text{return}} T_{\text{return}} + \dot{V}_{\text{makeup}} T_{\text{makeup}}}{\dot{V}_{\text{return}} + \dot{V}_{\text{makeup}}} \\ &= \frac{\left(60\frac{\text{gal}}{\text{min}}\right)(120.33\text{°F}) + \left(44\frac{\text{gal}}{\text{min}}\right)(53\text{°F})}{60\frac{\text{gal}}{\text{min}} + 44\frac{\text{gal}}{\text{min}}} \\ &= 91.8\text{°F}\end{aligned}$$

The answer is (A).

Why Other Options Are Wrong

(B) This answer results from neglecting to reduce the return flow before calculating the return water temperature.

(C) This answer results from neglecting to convert gallons to cubic feet when calculating interim temperatures.

(D) This answer results from considering only the return water volume in the denominator, rather than the total mixture volume flow.

Depth

SOLUTION 19

The total resistance of the skylight frame includes the frame's R-value plus the indoor and outdoor surface resistances.

$$R_{\text{total}} = R_{\text{outdoor air film}} + R_{\text{frame}} + R_{\text{indoor air film}}$$

The surface resistances are the reciprocals of the surface conductances.

$$\begin{aligned} R_{\text{total}} &= \frac{1}{h_{\text{outdoor}}} + R_{\text{frame}} + \frac{1}{h_{\text{indoor}}} \\ &= \frac{1}{6.0\ \frac{\text{Btu}}{\text{hr-ft}^2\text{-°F}}} + 2.0\ \frac{\text{hr-ft}^2\text{-°F}}{\text{Btu}} + \frac{1}{1.60\frac{\text{Btu}}{\text{hr-ft}^2\text{-°F}}} \\ &= 2.79\ \frac{\text{hr-ft}^2\text{-°F}}{\text{Btu}} \end{aligned}$$

The temperature drop between indoor and outdoor temperatures at a point in an assembly is proportional to the R-value at that point, relative to the total R-value.

$$\frac{\Delta T_{\text{component}}}{\Delta T_{\text{overall}}} = \frac{R_{\text{component}}}{R_{\text{overall}}}$$

$$\Delta T_{\text{component}} = \Delta T_{\text{overall}}\left(\frac{R_{\text{component}}}{R_{\text{overall}}}\right)$$

The indoor surface temperature of the frame is thus calculated by subtracting from room temperature the temperature drop due to the resistance of the indoor air film.

$$\begin{aligned} \Delta T_{\text{indoor air film}} &= \Delta T_{\text{total}}\left(\frac{R_{\text{indoor air film}}}{R_{\text{total}}}\right) \\ &= (72\text{°F} - (-60\text{°F}))\left(\frac{\frac{1}{1.60\ \frac{\text{Btu}}{\text{hr-ft}^2\text{-°F}}}}{2.79\ \frac{\text{hr-ft}^2\text{-°F}}{\text{Btu}}}\right) \\ &= 29.6\text{°F} \end{aligned}$$

$$\begin{aligned} T_{\text{surface}} &= T_{\text{room}} - \Delta T_{\text{indoor air film}} \\ &= 72\text{°F} - 29.6\text{°F} \\ &= 42.4\text{°F} \end{aligned}$$

To avoid condensation, the room air dew point must be less than 42.4°F. The corresponding relative humidity at room temperature is determined by locating 43°F on the saturation curve of the psychrometric chart and following a line of constant humidity to the right until it intercepts the 72°F dry-bulb temperature line. This intersection corresponds to approximately 35% relative humidity.

The answer is (C).

Why Other Options Are Wrong

(A) This incorrect answer results from assuming that the relevant dew point is the outdoor temperature, rather than the inside frame surface temperature.

(B) The incorrect answer of 20% relative humidity results from using the temperature difference across the air film (29.6°F) as the dew point, rather than the frame surface temperature.

(D) This incorrect answer results from confusing saturation relative humidity with dew point at a particular frame surface temperature.

SOLUTION 20

The process line between the two state points on the psychrometric chart is represented here with only relevant lines shown.

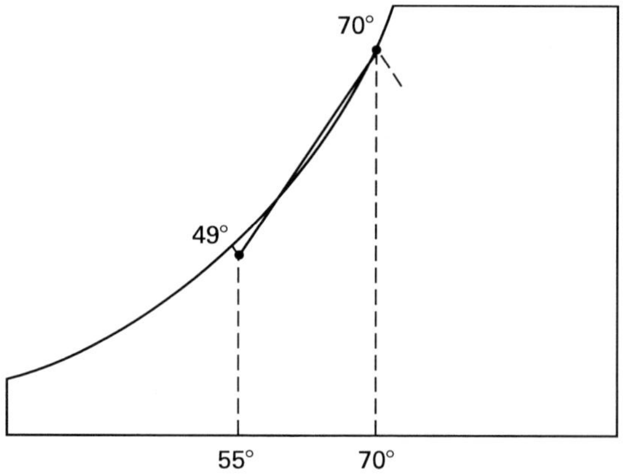

The process line crosses the curved saturation line. This shows the potential of this process to raise the moisture beyond saturation. The process thus has the potential to create fog, which consists of liquid droplets suspended in the atmosphere.

The answer is (C).

Why Other Options Are Wrong

(A) Adsorption is a process in which molecules of gas or liquid are extracted from an atmosphere. But it does not indicate suspended moisture droplets such as those that characterize fog.

(B) Condensation is a process that occurs when moisture comes out of phase by encountering colder objects.

(D) Sublimation indicates a change of phase, but from a solid directly to a gas without passing through a liquid phase.

SOLUTION 21

The efficiency of a cooling tower is defined as its range divided by the sum of its range and approach.

$$\eta_{\text{tower}} = \frac{\text{range}}{\text{range} + \text{approach}}$$

The range and approach of a cooling tower are defined as

$$\begin{aligned} \text{range} &= T_{\text{entering}} - T_{\text{leaving}} \\ \text{approach} &= T_{\text{leaving}} - T_{\text{wet-bulb}} \end{aligned}$$

Substituting the definition temperatures into the efficiency equation gives

$$\begin{aligned} \eta_{\text{tower}} &= \frac{T_{\text{entering}} - T_{\text{leaving}}}{(T_{\text{entering}} - T_{\text{leaving}}) + (T_{\text{leaving}} - T_{\text{wet-bulb}})} \\ &= \frac{T_{\text{entering}} - T_{\text{leaving}}}{T_{\text{entering}} - T_{\text{wet-bulb}}} \end{aligned}$$

Solving for the leaving water temperature,

$$\begin{aligned} T_{\text{leaving}} &= (1 - \eta_{\text{tower}})T_{\text{entering}} + \eta_{\text{tower}}T_{\text{wet-bulb}} \\ &= (1 - 0.52)(90°\text{F}) + (0.52)(75°\text{F}) \\ &= 82.2°\text{F} \quad (82°\text{F}) \end{aligned}$$

A simpler approach is to understand that the maximum theoretical cooling occurs if the 90°F incoming water can be cooled to the wet-bulb temperature of 75°F. A tower that is 52% efficient achieves 52% of the 15°F of cooling, which is 7.8°F. Subtracted from the 90°F incoming temperature, that results in a leaving water temperature of 82.2°F.

The answer is (C).

Why Other Options Are Wrong

(A) This answer would result if, in using the simplified approach described, the 7.8°F of cooling were subtracted from the 75°F wet-bulb temperature. Upon inspection, it should be recognized that the result is theoretically impossible.

(B) This incorrect answer could only be true for a 100% efficient cooling tower.

(D) This answer is incorrect due to transposing the efficiency and the complement of the efficiency when calculating the "leaving water temperature."

SOLUTION 22

It is not possible to cool the water below the air's wet-bulb temperature. Thus, the wet-bulb temperature, more than any other parameter, sets the absolute limit on the cooling tower's capacity.

The answer is (D).

Why Other Options Are Wrong

(A) This answer is incorrect. The approach is the difference between the leaving water temperature and the wet-bulb temperature of the incoming air. It has only an indirect influence on thermal performance.

(B) This answer is incorrect. The range is the difference between the incoming and desired outgoing water temperatures. While this has influence, it is nearly irrelevant if the wet-bulb temperature is high.

(C) This answer is incorrect. Cooling tower performance is only indirectly affected by outdoor dry-bulb temperature.

SOLUTION 23

The process is shown here with only the relevant lines of the psychrometric chart given.

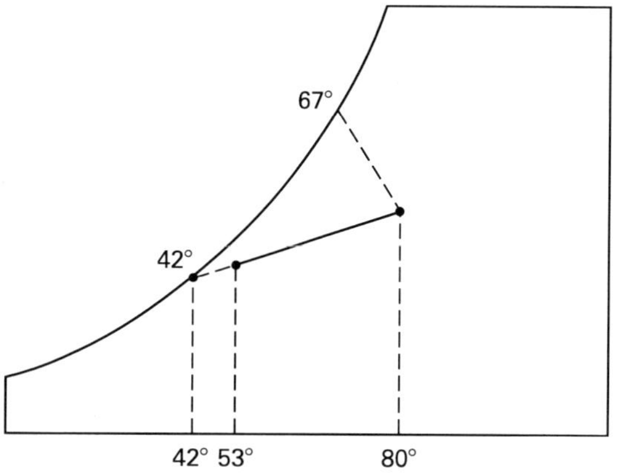

The reheat coil process is extraneous to the cooling coil efficiency concern, and so is not shown. The coil efficiency is

$$\eta = 1.0 - \text{BF}$$

The bypass factor is determined from the entering and exiting dry-bulb temperatures and the coil temperature, also called the apparatus dew point. The bypass factor is

$$\begin{aligned} \text{BF} &= \frac{T_{\text{db,out}} - T_{\text{coil}}}{T_{\text{db,in}} - T_{\text{coil}}} \\ &= \frac{53°\text{F} - 42°\text{F}}{80°\text{F} - 42°\text{F}} \\ &= 0.29 \end{aligned}$$

The coil efficiency is thus

$$\eta = 1.0 - 0.29$$
$$= 0.71 \quad (0.70)$$

The answer is (C).

Why Other Options Are Wrong

(A) This answer results from incorrectly identifying the bypass factor as the coil efficiency.

(B) This answer results from erroneously using the 66°F air leaving the reheat coil to determine the cooling coil efficiency.

(D) This incorrect answer describes a coil that is 100% efficient. If this were the case, the air leaving the coil would have a 42°F dry-bulb and a 42°F wet-bulb temperature.

SOLUTION 24

The bypass factor represents the fraction of air that passes through the coil without encountering the tubes or fins, thus leaving with its temperature unchanged. It is described quantitatively as

$$\text{BF} = \frac{T_{\text{db,out}} - T_{\text{coil}}}{T_{\text{db,in}} - T_{\text{coil}}}$$

Both $T_{\text{db,out}}$ and $T_{\text{db,in}}$ are dry-bulb temperatures. In this case the bypass factor is

$$\text{BF} = \frac{54°\text{F} - 42°\text{F}}{85°\text{F} - 42°\text{F}}$$
$$= 0.28$$

The equation for the bypass factor can be rearranged to solve for the temperature of the air leaving the coil at the second condition.

$$T_{\text{db,out}} = T_{\text{coil}} + (T_{\text{db,in}} - T_{\text{coil}})(\text{BF})$$
$$= 38°\text{F} + (80°\text{F} - 38°\text{F})(0.28)$$
$$= 49.8°\text{F} \quad (50°\text{F})$$

The answer is (C).

Why Other Options Are Wrong

(A) This incorrect answer results from noting that the incoming temperature is 5°F less, and the dew point is 4°F less, and reasoning that the outgoing temperature will be 9°F less.

(B) This incorrect answer results from guessing that the temperature drop will be a constant 31°F regardless of the other temperatures involved.

(D) This incorrect answer results from using the temperature of the entering chilled water, rather than the apparatus dew point (or overall coil temperature) to determine the bypass factor.

SOLUTION 25

Infinite variation over the controller range is accomplished with modulating control, which is also described as proportional, PI (proportional-integral), or PID (proportional-integral-derivative) action.

The answer is (C).

Why Other Options Are Wrong

(A) In two-position control, as with a simple heating thermostat, the control device can only be positioned to on or off. It does not allow infinite variation.

(B) Floating action lets the controller move back and forth between open and closed positions. It does not allow infinite variation.

(D) Fuzzy logic describes a more modern control algorithm that uses a series of if/then rules to approximate the way a person might hand-control the process. It does not describe this kind of action.

SOLUTION 26

All the heat removed by a refrigeration system, including the heat it introduces due to the work of the compressor, must ultimately be rejected to the outdoor environment by the condenser. So refrigeration effect *and* compressor work must be considered.

The answer is (C).

Why Other Options Are Wrong

(A) This answer is incorrect because subcooling is only part of the refrigeration effect.

(B) This answer is incorrect because it does not include the refrigeration effect.

(D) This answer is incorrect because it limits the answer to superheating, which is only a part of the refrigerating effect.

SOLUTION 27

The time constant represents the time required for the output of a system, forced by a step or impulse, to complete 63.2% of the total rise or decay.

In the situation described, 63.2% of the response time after activation of the control valve is

$$\begin{aligned}\tau &= 0.632t\\ &= (0.632)(50\ \text{sec})\\ &= 31.6\ \text{sec} \quad (32\ \text{sec})\end{aligned}$$

The answer is (B).

Why Other Options Are Wrong

(A) This incorrect answer represents 50% of the time it takes to achieve steady-state value, which is often defined as the time *delay*, not the time *constant*.

(C) This incorrect answer is the time it takes to achieve steady-state response. This is not the time constant.

(D) This answer is incorrect. There is no physical principle or definition in control response based upon a one-minute interval.

SOLUTION 28

The maximum flow will be achieved when the fan brake horsepower (BHP) equals the installed motor horsepower. According to the third fan law, the relationship between horsepower and rotational speed from an initial condition, 1, to final condition, 2, is

$$\frac{P_2}{P_1} = \left(\frac{n_2}{n_1}\right)^3$$

Since the airflow ratio is equal to the speed ratio, the final achievable air flow is

$$\begin{aligned}Q_2 &= Q_1 \sqrt[3]{\frac{P_2}{P_1}}\\ &= \left(35{,}000\ \frac{\text{ft}^3}{\text{min}}\right)\sqrt[3]{\frac{20.0\ \text{hp}}{16.2\ \text{hp}}}\\ &= 37{,}547\ \text{ft}^3/\text{min} \quad (38{,}000\ \text{ft}^3/\text{min})\end{aligned}$$

The answer is (B).

Why Other Options Are Wrong

(A) This incorrect answer results from confusing the motor horsepower and initial brake horsepower (BHP).

(C) This incorrect answer results from taking the square root of the horsepower ratio instead of the cube root.

(D) This incorrect answer results from multiplying the initial speed by the power ratio itself, rather than by the cube root of the power ratio.

SOLUTION 29

A pitot tube is normally connected to the manometer so that one side reads total pressure and the other side reads static pressure. The difference between the two columns of water gives velocity pressure. The velocity head in any fluid stream is given as

$$\begin{aligned}h_\text{v} &= \frac{\text{v}^2}{2g}\\ &= \left(\frac{\left(2500\ \frac{\text{ft}}{\text{min}}\right)^2\left(\frac{1\ \text{min}}{60\ \text{sec}}\right)^2}{(2)\left(32.2\ \frac{\text{ft}}{\text{sec}^2}\right)}\right)\left(12\ \frac{\text{in}}{\text{ft}}\right)\\ &\quad\times\left(0.075\ \frac{\text{lbm}}{\text{ft}^3\ \text{of air}}\right)\left(\frac{1\ \text{ft}^3\ \text{of water}}{62.34\ \text{lbm}}\right)\\ &= 0.3892\ \text{in of water} \quad (0.39\ \text{in wg})\end{aligned}$$

The answer is (B).

Why Other Options Are Wrong

(A) This answer results from neglecting to convert feet of water gage to inches of water gage.

(C) This answer results from calculating the velocity head term straight from the equation, neglecting all unit conversions.

(D) This answer results from neglecting to convert minutes to seconds in the nonstandard velocity term.

SOLUTION 30

For two dynamically similar (homologous) pumps, pump speed, impeller diameter, and total dynamic head are related by the equation

$$\frac{n_1 d_1}{\sqrt{h_1}} = \frac{n_2 d_2}{\sqrt{h_2}}$$

The speed of the manufacturer's pump, both as manufactured (pump 1) and after the impeller is trimmed

Depth

(pump 2), is 1750 rpm. Pump 1 has a diameter of 12.5 in and delivers 140 ft of total dynamic head. The diameter required to develop 115 ft at the same speed is

$$\begin{aligned} d_2 &= d_1 \frac{n_1\sqrt{h_2}}{n_2\sqrt{h_1}} \\ &= (12.5\ \text{in})\left(\frac{(1750\ \text{rpm})\sqrt{115\ \text{ft}}}{(1750\ \text{rpm})\sqrt{140\ \text{ft}}}\right) \\ &= 11.3\ \text{in} \quad (11\ \text{in}) \end{aligned}$$

The answer is (D).

Why Other Options Are Wrong

(A) This incorrect answer results when the ratio of pump dynamic heads is cubed.

(B) This incorrect answer results when the ratio of pump dynamic heads is squared.

(C) This incorrect answer results when the initial impeller diameter is multiplied by the simple ratio of pump dynamic heads.

SOLUTION 31

Centrifugal chillers are generally selected for their high coefficient of performance, good part-load performance, and cooling capacity. The economics of large-scale applications (such as a hospital) generally favor more efficient machines such as centrifugal chillers. In this case, a centrifugal chiller is also the only one of the types listed that can provide the needed capacity.

The answer is (C).

Why Other Options Are Wrong

(A) This answer is incorrect because reciprocating chillers are generally available only up to 450 tons.

(B) This answer is incorrect because scroll chillers are available only up to 80 tons.

(D) This answer is incorrect because screw chillers are available only up to 1000 tons.

SOLUTION 32

At the saturation temperature, gas solubility is near zero. Thus, the amount of steam required is that necessary to heat the condenser water to saturation temperature at 5 psig. This would be the product of the condensate's mass flow rate, specific heat, and temperature change.

$$Q_{\text{water}} = \dot{m}_{\text{water}} c_p \Delta T_{\text{water}}$$

The absolute pressure of the 5 psig tank is

$$\begin{aligned} p_{\text{tank}} &= p_{gage} + p_{\text{atm}} \\ &= 5\ \frac{\text{lbf}}{\text{in}^2} + 14.7\ \frac{\text{lbf}}{\text{in}^2} \\ &= 19.7\ \text{psia} \end{aligned}$$

The boiling temperature of water at 19.7 psia is 227°F. The heat required to bring the condensate to boiling is thus

$$\begin{aligned} Q_{\text{water}} &= \dot{m}_{\text{water}} c_p \Delta T_{\text{water}} \\ &= \left(1000\ \frac{\text{lbm}}{\text{hr}}\right)\left(1.0\ \frac{\text{Btu}}{\text{lbm-°F}}\right)(227°\text{F} - 160°\text{F}) \\ &= 67{,}000\ \text{Btu/hr} \end{aligned}$$

The steam required is this heating rate divided by the enthalpy change of the steam.

$$\dot{m}_{\text{steam}} = \frac{Q_{\text{water}}}{h_g - h_f}$$

From steam tables, the enthalpy of 30 psig (44.7 psia) steam at 350°F is

$$h_g = 1210.5\ \text{Btu/lbm}$$

The enthalpy of 227°F saturated liquid (original condensate and condensed steam) is

$$h_f = 195.4\ \text{Btu/lbm}$$

The steam needed to heat the condensate is

$$\begin{aligned} \dot{m}_{\text{steam}} &= \frac{67{,}000\ \frac{\text{Btu}}{\text{hr}}}{1210.5\ \frac{\text{Btu}}{\text{lbm}} - 195.4\ \frac{\text{Btu}}{\text{lbm}}} \\ &= 66.003\ \text{lbm/hr} \quad (66\ \text{lbm/hr}) \end{aligned}$$

The answer is (B).

Why Other Options Are Wrong

(A) This incorrect answer results from using 5 psig in the steam tables instead of converting to psia.

(C) This incorrect answer results from reading h_g at 30 psig steam at saturation, rather than at the 350°F superheated condition.

(D) This incorrect answer results from reading h_{fg} at 30 psig steam enthalpy at saturation, and using this value in place of h_g at 350°F superheated condition.

SOLUTION 33

The maximum rate of steam generation would be achieved by providing just the heat required to convert the water to saturated steam. This requires sensible heating of the hot water to the saturated liquid stage, as well as the heating to accomplish the phase change to saturated steam.

$$Q = \dot{m}c_p\Delta T + \dot{m}h_{fg} = \dot{m}(h_2 - h_1)$$

The mass flow rate is

$$\dot{m} = \frac{Q}{h_2 - h_1}$$

The absolute pressure of the saturated steam is the sum of the gage pressure and atmospheric pressure.

$$\begin{aligned} p_{\text{abs}} &= p_{\text{gage}} + p_{\text{atm}} \\ &= 15\ \frac{\text{lbf}}{\text{in}^2} + 14.7\ \frac{\text{lbf}}{\text{in}^2} \\ &= 29.7\ \text{psia} \quad (30\ \text{psia}) \end{aligned}$$

The enthalpy of water at 180°F and 1 atm is 148.04 Btu/lbm. The enthalpy of saturated steam at 29.7 psia (using 30 psia) is approximately 1164.1 Btu/lbm. The required mass flow rate is

$$\begin{aligned} \dot{m} &= \frac{Q}{h_2 - h_1} \\ &= \frac{(100\ \text{bhp})\left(\dfrac{33{,}475\ \dfrac{\text{Btu}}{\text{hr}}}{1\ \text{bhp}}\right)}{1164.1\ \dfrac{\text{Btu}}{\text{lbm}} - 148.04\ \dfrac{\text{Btu}}{\text{lbm}}} \\ &= 3294.6\ \text{lbm/hr} \quad (3300\ \text{lbm/hr}) \end{aligned}$$

The answer is (A).

Why Other Options Are Wrong

(B) This answer results from incorrectly taking the steam properties at 15 psia instead of 29.7 psia.

(C) This incorrect answer results from the common shortcut of ignoring sensible heating and taking h_{fg} to be 1000 Btu/lbm.

(D) This incorrect answer results from neglecting the sensible heating of the hot water to saturation temperature.

SOLUTION 34

The furnace has positive vent pressure and the steady-state efficiency is 81%, so the flue losses cannot be less than 19%. The unit cannot meet the efficiency threshold of Category IV, so it must be classed in Category III.

The answer is (C).

Why Other Options Are Wrong

(A) This answer is incorrect. The furnace must operate with a non-positive vent pressure to be in Category I.

(B) This answer is incorrect. The furnace must operate with a non-positive vent pressure to be in Category II.

(D) This answer is incorrect. The furnace's flue loss must be greater than 17% to be in Category IV.

SOLUTION 35

The relationship between the flow through a control valve and the pressure drop across it is shown by the nondimensionally homogeneous equation

$$Q_{\text{gpm}} = C_{\text{v}}\sqrt{\frac{\Delta p_{\text{psi}}}{\text{SG}}}$$

The required flow coefficient is thus given by the equation

$$C_{\text{v}} = \frac{Q_{\text{gpm}}}{\sqrt{\dfrac{\Delta p_{\text{psi}}}{\text{SG}}}}$$

The chilled water flow required to meet the cooling load is

$$\begin{aligned} Q_{\text{coil}} &= \dot{m}c_p(T_{\text{out}} - T_{\text{in}}) \\ &= \rho Q_{\text{gpm}}c_p(T_{\text{out}} - T_{\text{in}}) \\ Q_{\text{gpm}} &= \frac{Q_{\text{coil}}}{\rho c_p(T_{\text{out}} - T_{\text{in}})} \\ &= \frac{\left(650{,}000\ \dfrac{\text{Btu}}{\text{hr}}\right)\left(\dfrac{1\ \text{hr}}{60\ \text{min}}\right)}{\left(62.34\ \dfrac{\text{lbm}}{\text{ft}^3}\right)\left(\dfrac{1\ \text{ft}^3}{7.48\ \text{gal}}\right)} \\ &\quad \times\left(1.0\ \frac{\text{Btu}}{\text{lbm-°F}}\right)(54°\text{F} - 42°\text{F}) \\ &= 108.3\ \text{gpm} \end{aligned}$$

The maximum control valve pressure drop is

$$\Delta p_{\text{max}} = (0.25)\Delta p_{\text{total}}$$

$$= (0.25)(60 \text{ ft wg})\left(\frac{1 \frac{\text{lbf}}{\text{in}^2}}{2.31 \text{ ft wag}}\right)$$

$$= 6.5 \text{ psi}$$

Taking the specific gravity (SG) of the chilled water to be 1.0 (any difference from this will be too small to affect the answer), the dimensionless flow coefficient is found to be

$$C_v = \frac{108.3 \frac{\text{gal}}{\text{min}}}{\sqrt{\frac{6.5 \frac{\text{lbf}}{\text{in}^2}}{1.0}}}$$

$$= 42.48 \quad (42)$$

The answer is (C).

Why Other Options Are Wrong

(A) This incorrect answer is due to dividing by the whole pressure/specific gravity ratio instead of by its square root.

(B) This answer is incorrect because the fluid head was not converted to psi.

(D) This answer is incorrect due to failing to convert hours to minutes.

SOLUTION 36

The third affinity law relates pump brake horsepower and speed, from an initial condition 1 to a final condition 2, as

$$\frac{P_2}{P_1} = \left(\frac{n_2}{n_1}\right)^3$$

The first affinity law indicates that the ratios of flow and pump speed are proportional, allowing the use of either ratio interchangeably.

$$\frac{Q_2}{Q_1} = \frac{n_2}{n_1}$$

$$\frac{P_2}{P_1} = \left(\frac{Q_2}{Q_1}\right)^3$$

The power required at the new speed is therefore

$$P_2 = P_1\left(\frac{Q_2}{Q_1}\right)^3$$

$$= (86.6 \text{ hp})\left(\frac{1050 \frac{\text{gal}}{\text{min}}}{1400 \frac{\text{gal}}{\text{min}}}\right)^3$$

$$= 36.5 \text{ hp}$$

A 40 hp motor is the next available standard size.

The answer is (B).

Why Other Options Are Wrong

(A) Though close, this pump would be slightly undersized. Nearly as important, this is a nonstandard motor size that would not be readily available.

(C) This incorrect answer results from taking the square of the ratio rather than the cube.

(D) This answer results if the horsepower is incorrectly assumed to be proportional to flow.

SOLUTION 37

The reversing valve reverses the direction of refrigerant vapor flow between the evaporator, compressor, and condenser. When the path of the refrigerant is reversed, the evaporator becomes the condenser and the condenser becomes the evaporator. In this way, the heat pump can heat a space by absorbing energy from a medium (such as outside air) or cool a space by rejecting heat to the same medium.

The answer is (C).

Why Other Options Are Wrong

(A) This incorrect answer describes a valve, but not a reversing valve.

(B) This incorrect answer does not describe a reversing valve, and is also incorrect because it would not be possible to reverse the flow through a coil with a two-way valve if the pump pressure continued to be exerted from the same side.

(D) This answer is incorrect. There is no common usage of the term "reversing valve" in the automatic temperature control industry.

SOLUTION 38

Oil return to the compressor is important to avoid starving the crankcase of adequate oil for compressor lubrication. But oil should be returned directly to the crankcase via refrigerant return to the suction side of the compressor, not the discharge side.

Neither refrigerant head nor a minimum velocity is a concern. The energy savings involved could range from trivial to nonexistant.

The answer is (D).

Why Other Options Are Wrong

(A) This answer is incorrect. The refrigerant is in the form of "hot gas." There should be very little refrigerant head going into a condenser.

(B) This answer is incorrect. There is no minimum velocity criterion for refrigerant entering a condenser.

(C) This answer is incorrect. All things being equal, non-pitched pipe would not necessarily result in a *net* increase of elevation through which refrigeration would have to be moved. In any case, energy savings, if any, would be insignificant.

SOLUTION 39

Piping schedule numbers indicate the approximate value of the expression p/S (the ratio of service pressure to allowable stress, both expressed in units of lbf/in^2) multiplied by 1000. The schedule number itself is dimensionless.

The answer is (B).

Why Other Options Are Wrong

(A) While maximum velocity guidelines exist, they are not part of a pipe's schedule designation.

(C) The schedule number does not refer to roughness, which is usually measured in feet, not thousandths of an inch.

(D) The Brinell hardness number measures the relative hardness of metals and alloys, but the schedule number does not refer to this.

SOLUTION 40

The net positive suction head available is

$$\text{NPSHA} = h_{\text{atm}} + h_z - h_{\text{vp}} + h_f$$

Atmospheric pressure, static, and friction head are given. To determine the vapor pressure of the water, take the saturation pressure of 0.698133 psi for water at 90°F from steam tables.

The net positive suction head available is

$$\begin{aligned}\text{NPSHA} &= \left(11.3\ \frac{\text{lbf}}{\text{in}^2}\right)\left(\frac{2.31\ \text{ft}}{1\ \frac{\text{lbf}}{\text{in}^2}}\right) + (-6.0\ \text{ft}) \\ &\quad - \left(0.698133\ \frac{\text{lbf}}{\text{in}^2}\right)\left(\frac{2.31\ \text{ft}}{1\ \frac{\text{lbf}}{\text{in}^2}}\right) - 8.0\ \text{ft} \\ &= 10.49\ \text{ft} \quad (10\ \text{ft})\end{aligned}$$

The answer is (B).

Why Other Options Are Wrong

(A) This incorrect answer results from failing to convert psia to feet of head.

(C) This incorrect answer results from neglecting to include an allowance for the liquid vapor pressure at the indicated temperature.

(D) This incorrect answer results from omitting the suction lift, h_z, from the calculation.

SOLUTION 41

ASHRAE Standard 55, *Thermal Environmental Conditions for Human Occupancy,* describes environmental criteria for occupied spaces, but not the performance requirement of the systems and components that condition the space.

The answer is (B).

Why Other Options Are Wrong

(A) This answer is incorrect. ANSI/ASHRAE/IESNA Standard 90.1 describes energy efficiency (performance) standards not only for HVAC equipment, but for building envelope and lighting systems as well.

(C) This answer is incorrect. ANSI Z21.10/CSA 4 describes the test procedures for water heating equipment performance testing.

(D) This answer is incorrect. This publication of the Cooling Tower Institute (CTI) outlines test procedures for establishing the performance of evaporative cooling towers.

SOLUTION 42

The relation between sensible load and the temperature rise of the airflow is

$$\begin{aligned} Q_{\text{sensible}} &= \dot{m}c_p(T_{\text{air,out}} - T_{\text{air,in}}) \\ &= \dot{V}\rho c_p(T_{\text{air,out}} - T_{\text{air,in}}) \end{aligned}$$

The incoming air is at outdoor conditions, and the outgoing air temperature is that which must not exceed the transformer vault maximum temperature.

Efficiency in this case is that part of the transformer power that is not wasted as heat. The wasted power is the heat that must be removed from the vault. The needed quantity of air, then, is not for the full heat transfer rate of 5 MW but is for the wasted part of that rate.

The required volume flow rate is thus

$$\begin{aligned} \dot{V} &= \frac{Q_{\text{sensible}}(1-\eta)}{\rho c_p(T_{\text{air,out}} - T_{\text{air,in}})} \\ &= \left(\frac{(5\times 10^6\ \text{W})\left(\dfrac{3.412\ \dfrac{\text{Btu}}{\text{hr}}}{1\ \text{W}}\right)(1-0.97)}{\left(0.075\ \dfrac{\text{lbm}}{\text{ft}^3}\right)\left(0.24\ \dfrac{\text{Btu}}{\text{lbm-°F}}\right)(135°\text{F} - 105°\text{F})}\right) \\ &\quad \times\left(\frac{1\ \text{hr}}{60\ \text{min}}\right) \\ &= 15{,}796\ \text{ft}^3/\text{min} \quad (16{,}000\ \text{ft}^3/\text{min}) \end{aligned}$$

The answer is (C).

Why Other Options Are Wrong

(A) This incorrect answer results from not converting watts to Btu/hr.

(B) This incorrect answer is based on the temperature of the nearby building rather than the outdoor air as specified.

(D) This incorrect answer results from assuming that all the power of the transformer is rejected as heat to the vault, rather than just the losses.

SOLUTION 43

Table 6A of the Uniform Mechanical Code contains the minimum gage requirements for rectangular ductwork to provide adequate structural integrity and avoid "oil-canning." The gage number *decreases* as the dimension of the longest side *increases*. For example, duct dimensions up through 12 in require 26 gage ductwork, which is approximately 0.019 in thick. Duct dimensions from 13 to 18 in require 24 gage ductwork, which is approximately 0.024 in thick.

The answer is (C).

Why Other Options Are Wrong

(A) If the gage increased with pressure, the sheet metal would get *thinner* with higher pressure.

(B) If the gage increased with maximum dimension, then the longer the span between corners, the *thinner* the ductwork would be.

(D) If the gage decreased with dimension, it would at least be getting thicker. But the shorter dimension is not the concern ("weak link") of the code in this situation. The longer side must be used to set the gage for the entire duct.

SOLUTION 44

Refrigerants are classified by ANSI/ASHRAE Standard 34 into safety groups as shown.

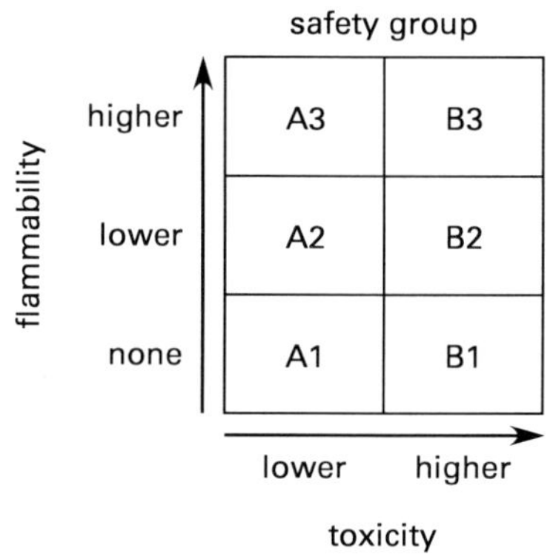

The answer is (B).

Why Other Options Are Wrong

(A) This answer is incorrect. 1 stands for the lowest level of flammability.

(C) This answer is incorrect. 2 stands for a lower level of flammability than 3.

(D) This answer is incorrect. C is not a valid designation.

SOLUTION 45

ASHRAE publishes comfort criteria as a region within the "comfort chart." The chart's coordinates are humidity and operative temperature. Numerically, operative temperature is the average of the air temperature and mean radiant temperature, weighted by the convection and radiation heat transfer coefficients (h_c and h_r). Operative temperature is stated in equation form as

$$T_o = \frac{h_c T_a + h_r T_r}{h_c + h_r}$$

The answer is (B).

Why Other Options Are Wrong

(A) This answer is incorrect. Dry-bulb temperature is a parameter used in calculating operative temperature.

(C) This answer is incorrect. Mean radiant temperature is a parameter used in calculating operative temperature.

(D) This answer is incorrect. Air velocity is a parameter used in calculating operative temperature.

SOLUTION 46

According to ASHRAE Standard 62.1 Table 6-1, the required outside air rate by people, R_p, is 5 ft^3/min per person, and the required outside air rate by area, R_a, is 0.06 ft^3/min per ft^2. Since the zone population, P_z, is 85 persons and the zone floor area, A_z, is 12,000 ft^3, the total breathing zone outside air rate, V_{bz}, required for the office space is

$$\begin{aligned} V_{bz} &= R_p P_z + R_a A_z \\ &= \left(5\;\frac{\frac{\text{ft}^3}{\text{min}}}{\text{person}}\right)(85\text{ persons}) \\ &\quad + \left(\frac{0.06\;\frac{\text{ft}^3}{\text{min}}}{1\text{ ft}^2}\right)(12{,}000\text{ ft}^2) \\ &= 1145\text{ ft}^3/\text{min} \quad (1100\text{ cfm}) \end{aligned}$$

The answer is (A).

Why Other Options Are Wrong

(B) This incorrect answer is based on a common rule of thumb of 0.15 cfm per square foot of floor space.

(C) This incorrect answer results from relying on a common rule of thumb that outside air should be 15% of the total supply air.

(D) This incorrect answer results from confusing total supply air with the ventilation air component of supply air.

SOLUTION 47

The required percentage of outside air is the total outside air divided by the total supply air.

$$\text{oa}_{\%} = \frac{\dot{V}_{\text{oa}}}{\dot{V}_{\text{sa}}} \times 100\%$$

When the occupancy is not known or specified, the standard recommends a density for supermarket occupancies of 8 people per 1000 ft^2.

$$\begin{aligned} \text{no. of persons} &= \text{rate} \times \text{area} \\ &= \left(\frac{8\text{ persons}}{1000\text{ ft}^2}\right)(11{,}500\text{ ft}^2) \\ &= 92\text{ persons} \end{aligned}$$

The total outside air is the number of people times the prescribed outside air per person.

$$\begin{aligned} \dot{V}_{\text{oa}} &= n_{\text{persons}} \dot{V}_{\text{per person}} \\ &= (92\text{ persons})\left(\frac{7.5\;\frac{\text{ft}^3}{\text{min}}}{1\text{ person}}\right) \\ &\quad + (11{,}500\text{ ft}^2)\left(0.06\;\frac{\frac{\text{ft}^3}{\text{min}}}{\text{ft}^2}\right) \\ &= 1380\text{ ft}^3/\text{min} \end{aligned}$$

Thus,

$$\begin{aligned} \text{oa}_{\%} &= \left(\frac{1380\;\frac{\text{ft}^3}{\text{min}}}{12{,}000\;\frac{\text{ft}^3}{\text{min}}}\right) \times 100\% \\ &= 11.5\% \quad (12\%) \end{aligned}$$

The answer is (B).

Why Other Options Are Wrong

(A) This answer is incorrect. 10% is often used as a rule of thumb, but in this case it fails to meet ASHRAE standards.

(C) This answer is incorrect. 25% is often used as a rule of thumb, but in this case it gives a result more than twice the minimum needed by ASHRAE standards.

(D) This answer is incorrect. 33% is often used as a rule of thumb, but in this case it gives a result three times the minimum needed by ASHRAE standards.

SOLUTION 48

The characteristic dimension for a high sidewall grille is the distance from the grille to the opposite wall, which in this case is 30 ft. The throw of an air terminal device is the distance traveled by the time that velocity is reduced to 50 ft/min. This characteristic has been correlated with characteristic room dimensions in the ASHRAE Fundamentals Handbook. For the case of high sidewall grilles, the maximum ADPI is achieved for the given load densities when

$$\frac{T_{50}}{\mathrm{L}} = 1.5$$

The correct throw is

$$\begin{aligned} T_{50} &= 1.5L \\ &= (1.5)(30\ \text{ft}) \\ &= 45\ \text{ft} \end{aligned}$$

The answer is (D).

Why Other Options Are Wrong

(A) This incorrect answer results from confusing the load density and the throw.

(B) This incorrect answer results from using the old rule of thumb, "throw equals 3/4 of the distance to the nearest wall." This rule will not necessarily achieve acceptable ADPI levels.

(C) This incorrect answer results from confusing the characteristic dimension, L, with the throw.

SOLUTION 49

The relationship between sensible load, airflow, and temperature difference in a room is

$$\begin{aligned} Q_{\text{sensible}} &= \dot{m}c_p\Delta T \\ &= \rho\dot{V}c_p\Delta T \end{aligned}$$

The required volume flow rate is thus

$$\begin{aligned} \dot{V} &= \frac{Q_{\text{sensible}}}{\rho c_p \Delta T} \\ &= \frac{10{,}000\ \frac{\text{Btu}}{\text{hr}}}{\left(0.075\ \frac{\text{lbm}}{\text{ft}^3}\right)\left(0.24\ \frac{\text{Btu}}{\text{lbm-°F}}\right)(20°\text{F})\left(60\ \frac{\text{min}}{\text{hr}}\right)} \\ &= 460\ \text{ft}^3/\text{min} \end{aligned}$$

The closest entry in the "airflow" column of the table is 462 cfm.

Manufacturers typically provide throws for 50 ft/min, 100 ft/min, and 150 ft/min terminal velocities, as the table indicates. The ASHRAE air diffusion performance index (ADPI) method is specifically referenced to throws associated with 50 ft/min terminal velocities. In the diffuser performance data table, the correct throw required to maximize the ADPI is shown in the row for an airflow of 462 cfm and in the column for a throw of 50 fpm.

$$T_{50} = 8.0\ \text{ft}$$

The answer is (D).

Why Other Options Are Wrong

(A) This is the throw corresponding to a diffuser neck velocity of 450 ft/min, instead of a diffuser airflow of 450 cfm.

(B) This incorrect answer results from assuming a reference terminal velocity of 150 ft/min.

(C) This incorrect answer results from assuming a terminal velocity of 100 ft/min.

SOLUTION 50

The total pressure of the supply air is the sum of the static pressure and the velocity pressure.

$$\text{TP} = \text{SP} + \text{VP}$$

The density of air is approximately 0.075 lbm/ft^3. The initial velocity pressure is given by

$$\begin{aligned} VP_{\text{initial}} &= \frac{\rho v_{\text{initial}}^2}{2g_c} \\ &= \left(\frac{\left(0.075\ \frac{\text{lbm}}{\text{ft}^3}\right)\left(25\ \frac{\text{ft}}{\text{sec}}\right)^2}{(2)\left(32.2\ \frac{\text{ft-lbm}}{\text{lbf-sec}^2}\right)}\right) \\ &\quad \times \left(\frac{27.7\ \text{in wg}}{1\ \frac{\text{lbf}}{\text{in}^2}}\right)\left(\frac{1\ \text{ft}}{12\ \text{in}}\right)^2 \\ &= 0.14\ \text{in wg} \end{aligned}$$

So the total pressure of the CRAC unit discharge air is

$$\begin{aligned} TP &= 0.75\ \text{in wg} + 0.14\ \text{in wg} \\ &= 0.89\ \text{in wg} \end{aligned}$$

The static pressure at any point in the stream flowing away from the CRAC unit is the difference between the initial total pressure and the velocity pressure at that point.

$$SP = TP - VP$$

The air velocity at a distance of 8 ft from the unit can be determined from continuity.

$$v = \frac{\dot{V}}{A}$$

The cross-sectional area 8 ft from the unit is the product of plenum height and the circumference of the circle whose radius is 8 ft.

$$\begin{aligned} A &= hC = h2\pi r \\ &= (1.5\ \text{ft})2\pi(8\ \text{ft}) \\ &= 75.4\ \text{ft}^2 \end{aligned}$$

The air velocity 8 ft from the unit is

$$\begin{aligned} v_{8\,\text{ft}} &= \frac{\left(40{,}000\ \frac{\text{ft}^3}{\text{min}}\right)\left(\frac{1\ \text{min}}{60\ \text{sec}}\right)}{75.4\ \text{ft}^2} \\ &= 8.84\ \text{ft/sec} \end{aligned}$$

The static pressure 8 ft from the unit is

$$\begin{aligned} SP_{8\,\text{ft}} &= TP - VP_{8\,\text{ft}} = TP - \frac{\rho v_{8\,\text{ft}}^2}{2g_c} \\ &= 0.89\ \text{in wg} \\ &\quad - \left(\frac{\left(0.075\ \frac{\text{lbm}}{\text{ft}^3}\right)\left(8.84\ \frac{\text{ft}}{\text{sec}}\right)^2}{(2)\left(32.2\ \frac{\text{ft-lbm}}{\text{lbf-sec}^2}\right)}\right) \\ &\quad \times \left(\frac{27.7\ \text{in wg}}{1\ \frac{\text{lbf}}{\text{in}^2}}\right)\left(\frac{1\ \text{ft}}{12\ \text{in}}\right)^2 \\ &= 0.87\ \text{in wg} \end{aligned}$$

The answer is (C).

Why Other Options Are Wrong

(A) This incorrect answer results from subtracting the velocity pressure from the static pressure, rather than adding the two.

(B) This incorrect answer results from omitting the conversion from inches of water gage to pounds per square inch.

(D) This incorrect answer results from neglecting to square the velocity term in the final velocity pressure term.

SOLUTION 51

The equivalent diameter of a rectangular duct with dimensions a and b is

$$d_{\text{eq}} = \frac{1.30(ab)^{5/8}}{(a+b)^{1/4}}$$

The equivalent diameter of the given duct is

$$\begin{aligned} d_{\text{eq}} &= \frac{(1.30)\big((24\ \text{in})(36\ \text{in})\big)^{5/8}}{(24\ \text{in} + 36\ \text{in})^{1/4}} \\ &= 31.97\ \text{in} \quad (32\ \text{in}) \end{aligned}$$

The answer is (C).

Why Other Options Are Wrong

(A) This incorrect answer results from taking the product of the duct dimensions in the denominator instead of their sum.

Depth

(B) This incorrect answer is the hydraulic diameter, not the equivalent diameter.

(D) This incorrect answer results from calculating the equivalent area geometrically.

SOLUTION 52

The equal friction method is so called because the designer first determines a friction rate per unit of length, usually in inches of water gage per 100 ft of duct. Ductwork is then sized for required airflow rates so as to achieve the prescribed unit friction rate.

The answer is (C).

Why Other Options Are Wrong

(A) This incorrect answer describes the principle used for the static regain method.

(B) This incorrect answer describes the principle used for the ASHRAE T-method.

(D) This incorrect answer does not describe a duct design procedure.

SOLUTION 53

The dynamic loss through a duct fitting (in inches of water gage) is calculated as

$$\Delta p_f = C(\text{VP})$$

The velocity pressure is taken at the entry to the elbow and is given by

$$\text{VP} = \frac{\rho \text{v}^2}{2g_c}$$

The velocity at the fitting can be determined from continuity.

$$\text{v}_f = \frac{\dot{V}_f}{A_f}$$

The volume flow rate at the fitting is equal to the volume flow rate at the coil.

$$\begin{aligned} \dot{V}_f &= \dot{V}_c = \text{v}_c A_c \\ &= \left(500\ \frac{\text{ft}}{\text{min}}\right)(4\ \text{ft})(4\ \text{ft}) \\ &= 8000\ \text{ft}^3/\text{min} \end{aligned}$$

The velocity at the fitting is then

$$\begin{aligned} \text{v}_f &= \frac{\dot{V}_f}{A_f} \\ &= \frac{\left(8000\ \frac{\text{ft}^3}{\text{min}}\right)\left(\frac{1\ \text{min}}{60\ \text{sec}}\right)}{4\ \text{ft}^2} \\ &= 33.33\ \text{ft/sec} \end{aligned}$$

The fitting loss is

$$\begin{aligned} \Delta p &= C(\text{VP}) \\ &= C\left(\frac{\rho \text{v}^2}{2g_c}\right) \\ &= (1.4)\left(\frac{\left(0.075\ \frac{\text{lbm}}{\text{ft}^3}\right)\left(33.33\ \frac{\text{ft}}{\text{sec}}\right)^2}{(2)\left(32.2\ \frac{\text{ft-lbm}}{\text{lbf-sec}^2}\right)}\right) \\ &\quad \times \left(\frac{27.7\ \text{in wg}}{1\ \frac{\text{lbf}}{\text{in}^2}}\right)\left(\frac{1\ \text{ft}}{12\ \text{in}}\right)^2 \\ &= 0.3478\ \text{in wg} \quad (0.35\ \text{in wg}) \end{aligned}$$

The answer is (D).

Why Other Options Are Wrong

(A) This incorrect answer results from neglecting to square the velocity term.

(B) This answer results from incorrectly using the velocity through the cooling coil rather than through the elbow.

(C) This incorrect answer is the velocity pressure. The fitting coefficient has been neglected.

SOLUTION 54

The relationship between static pressure drop through any duct fitting and velocity pressure is

$$\Delta p_{\text{static}} = C(\text{VP})$$

The relationship between volume of airflow, duct velocity, and duct cross-sectional area is given by the continuity equation

$$\dot{V} = \text{v}A$$

The cross-sectional area of a 32 in by 40 in duct is

$$\begin{aligned} A &= hw \\ &= (32\text{ in})(40\text{ in})\left(\frac{1\text{ ft}}{12\text{ in}}\right)^2 \\ &= 8.889\text{ ft}^2 \end{aligned}$$

An airflow of 23,000 ft^3/min in the duct will have an average duct velocity of

$$\begin{aligned} \text{v} &= \frac{\dot{V}}{A} \\ &= \frac{\left(23{,}000\ \frac{\text{ft}^3}{\text{min}}\right)\left(\frac{1\text{ min}}{60\text{ sec}}\right)}{8.889\text{ ft}^2} \\ &= 43.12\text{ ft/sec} \end{aligned}$$

The velocity pressure (in inches of water gage) is

$$\begin{aligned} \text{VP} &= \frac{\rho \text{v}^2}{2g_c} \\ &= \left(\frac{\left(0.075\ \frac{\text{lbm}}{\text{ft}^3}\right)\left(43.12\ \frac{\text{ft}}{\text{sec}}\right)^2}{(2)\left(32.2\ \frac{\text{ft-lbm}}{\text{lbf-sec}^2}\right)}\right) \\ &\quad \times \left(\frac{27.7\text{ in wg}}{1\ \frac{\text{lbf}}{\text{in}^2}}\right)\left(\frac{1\text{ ft}}{12\text{ in}}\right)^2 \\ &= 0.417\text{ in wg} \end{aligned}$$

The height-to-width ratio of the duct is

$$\begin{aligned} \frac{h}{w} &= \frac{32\text{ in}}{40\text{ in}} \\ &= 0.80 \end{aligned}$$

Interpolating from the table, a height/width ratio of 0.80 and an angle of 35° gives an approximate fitting coefficient of 6.15.

$$\begin{aligned} \Delta p_{\text{static}} &= Cp_{\text{v}} \\ &= (6.15)(0.417\text{ in wg}) \\ &= 2.56\text{ in wg} \end{aligned}$$

The answer is (D).

Why Other Options Are Wrong

(A) This incorrect answer results from neglecting to square the velocity term when calculating the velocity pressure.

(B) This incorrect answer results from omitting the conversion from feet to inches.

(C) This incorrect answer results from leaving off the final step of conversion of velocity pressure to static pressure.

SOLUTION 55

The annual cost savings from this measure is the product of the savings in energy usage and the energy unit cost.

$$A_{\text{saved}} = E_{\text{saved}}C$$

The energy savings resulting from reduced pump runtime is the product of the power draw and the resulting reduction in number of hours of operation.

$$E_{\text{saved}} = Pt$$

For single phase motors, power in watts (volt-amps) is determined as the simple product of voltage and amperage.

$$P = VI$$

The power drawn by a three-phase electric motor where current is given for only one "leg" must be adjusted by the square root of 3, as well as by the power factor.

$$\begin{aligned} P &= VI\sqrt{3}(\text{PF}) = (460\text{ V})(8\text{ A})\sqrt{3}(0.88) \\ &= 5.61\text{ kW} \end{aligned}$$

The reduction in annual hours of pump operation is

$$t = \left(12\ \frac{\text{hr}}{\text{day}}\right)\left(365\ \frac{\text{day}}{\text{yr}}\right) = 4380\text{ hr/yr}$$

The annual energy savings is

$$E_{\text{saved}} = (5.61\text{ kW})\left(4380\ \frac{\text{hr}}{\text{yr}}\right) = 24{,}570\text{ kW-hr/yr}$$

The annual utility cost savings is

$$\begin{aligned} A_{\text{saved}} &= \left(24{,}570\ \frac{\text{kW-hr}}{\text{yr}}\right)\left(\frac{\$0.08}{\text{kW-hr}}\right) \\ &= \$1966/\text{yr} \quad (\$2000/\text{yr}) \end{aligned}$$

The answer is (B).

Depth

Why Other Options Are Wrong

(A) This incorrect answer results when the square root of 3 is omitted from the calculation. This and the power factor are both necessary to account for the three-phase power supplied.

(C) This incorrect answer results from trying to incorporate the pump efficiency into the estimate of power draw. The efficiency is already accounted for in the actual voltage and amperage data provided.

(D) This answer results from erroneously attributing monthly demand savings to this measure in addition to the energy savings. There will not likely be demand savings, because peak demand rates are normally charged based on the greatest demand for a short period of time, generally 15 minutes to one hour. The system still runs 12 hr per day, so it will almost certainly be operating during the period of peak demand for the building.

SOLUTION 56

According to ASHRAE Standard 62.1, the default outside air requirement for a school classroom (ages 5–8) is 10 ft^3/min per person plus 0.12 ft^3/min/ft^2. The default density is 25 persons per 1000 ft^2. The total outside air requirement for this room, then, is

$$\begin{aligned}\dot{V}_{\text{oa}} &= (1240\ \text{ft}^2)\left(\frac{25\ \text{persons}}{1000\ \text{ft}^3}\right)\left(\frac{10\ \frac{\text{ft}^3}{\text{min}}}{1\ \text{person}}\right) \\ &\quad + (1240\ \text{ft}^2)\left(0.12\ \frac{\frac{\text{ft}^3}{\text{min}}}{\text{ft}^2}\right) \\ &= 459\ \text{ft}^3/\text{min}\end{aligned}$$

The fraction of outside air required by the ventilation standard is

$$N_{\text{oa}} = \frac{\dot{V}_{\text{oa}}}{\dot{V}_{\text{total}}} = \frac{459\ \frac{\text{ft}^3}{\text{min}}}{1000\ \frac{\text{ft}^3}{\text{min}}} = 0.459$$

The mixed-air temperature entering the heat pump *without* auxiliary heating is

$$\begin{aligned}T_{\text{mixed}} &= N_{\text{oa}}T_{\text{oa}} + (1 - N_{\text{oa}})T_{\text{return}} \\ &= (0.459)(5^\circ\text{F}) + (1 - 0.459)(70^\circ\text{F}) \\ &= 40.2^\circ\text{F}\end{aligned}$$

This is below 45°F, so auxiliary heating will be necessary. The minimum electric heating capacity required is that which can preheat 1000 ft^3/min of air from 40.1°F to 45°F. The required minimum is

$$\begin{aligned}Q_{\text{sensible}} &= \dot{m}c_p(T_{\text{min}} - T_{\text{mixed}}) = \dot{V}\rho c_p(T_{\text{min}} - T_{\text{mixed}}) \\ &= \left(1000\ \frac{\text{ft}^3}{\text{min}}\right)\left(0.078\ \frac{\text{lbm}}{\text{ft}^3}\right)\left(0.24\ \frac{\text{Btu}}{\text{lbm-}^\circ\text{F}}\right) \\ &\quad \times(45^\circ\text{F} - 40.1^\circ\text{F})\left(60\ \frac{\text{min}}{\text{hr}}\right)\left(\frac{1\ \text{kW}}{3412\ \frac{\text{Btu}}{\text{hr}}}\right) \\ &= 1.613\ \text{kW} \quad (1.6\ \text{kW})\end{aligned}$$

The answer is (B).

Why Other Options Are Wrong

(A) This answer results from incorrectly estimating the ventilation requirement so that an entering temperature greater than 45°F is calculated.

(C) This answer results from calculating the heating necessary to raise the ventilation air temperature from 5°F to 45°F and failing to include the effect of the classroom return air.

(D) This incorrect answer results from calculating the heating necessary to raise all the air from 5°F to 45°F, neglecting that the classroom return air is at 70°F, not 5°F.

SOLUTION 57

From the heating capacity ratings given, the capacity of the heat pump at 10°F is approximately 29,000 Btu/hr. The auxiliary heating requirement is the load minus the heat pump capacity at the design temperature.

$$\begin{aligned}Q_{\text{aux}} &= Q_{\text{load}} - Q_{\text{cap}} \\ &= \left(55{,}000\ \frac{\text{Btu}}{\text{hr}} - 29{,}000\ \frac{\text{Btu}}{\text{hr}}\right)\left(\frac{1\ \text{kW}}{3412\ \frac{\text{Btu}}{\text{hr}}}\right) \\ &= 7.6\ \text{kW}\end{aligned}$$

The answer is (B).

Why Other Options Are Wrong

(A) This answer results from incorrectly using the 47°F standard ARI rating condition, rather than the actual performance at design conditions. The 60,000 Btu/hr capacity is greater than the 55,000 Btu/hr load.

(C) This incorrect answer results from converting the entire heating load directly to kilowatts, rather than that portion of the load unmet by the heat pump.

(D) This incorrect answer results from stating the answer in kBtu per hour ("MBH" in common industry terms) instead of converting to kilowatts.

SOLUTION 58

There are two acceptable procedures for achieving acceptable air quality, according to ANSI/ASHRAE Standard 62.1: the indoor air quality procedure and the ventilation rate procedure. The ventilation rate procedure prescribes the required delivery rate of outdoor air. If air contaminants in the local community do not exceed published values, establishing the proper rate for the space to be conditioned is all that is required.

The answer is (D).

Why Other Options Are Wrong

(A) No monitoring is required by the ventilation rate procedure, so this answer is incorrect.

(B) Air treatment is required only if *outdoor* air is unacceptable, so this answer is incorrect.

(C) It is the indoor air quality procedure, not the ventilation rate procedure, that requires the specification of acceptable indoor concentration of contaminants.

SOLUTION 59

ASHRAE Standard 84 defines effectiveness as

$$\epsilon = \frac{\text{actual transfer (of moisture or energy)}}{\text{maximum possible transfer between airstreams}}$$

The answer is (D).

Why Other Options Are Wrong

(A) This answer is incorrect. The Carnot cycle has nothing to do with heat transfer effectiveness.

(B) This answer is incorrect. The Carnot cycle has nothing to do with heat transfer effectiveness.

(C) This incorrect answer limits the definition to sensible heat recovery, whereas the ASHRAE definition does not.

SOLUTION 60

The transmissibility, T, is related to the square of the ratio of the disturbing frequency, f_d, to the system's natural frequency, f_n.

$$T = \left| \frac{1}{1 - \left(\frac{f_d}{f_n}\right)^2} \right|$$

The natural frequency is found from the static deflection.

$$\begin{aligned} f_n &= \frac{\omega}{2\pi} = \frac{1}{2\pi}\sqrt{\frac{g}{\delta_{\text{st}}}} \\ &= \frac{1}{2\pi}\sqrt{\frac{\left(32.2\ \frac{\text{ft}}{\text{sec}^2}\right)\left(12\ \frac{\text{in}}{\text{ft}}\right)}{1.25\ \text{in}}} \\ &= 2.80\ \text{Hz} \end{aligned}$$

The frequency of the rotating equipment is 700 rpm (cycles/min).

$$\begin{aligned} f_d &= \frac{700\ \frac{\text{cycles}}{\text{min}}}{60\ \frac{\text{sec}}{\text{min}}} \\ &= 11.67\ \text{cycles/sec} \quad (11.67\ \text{Hz}) \end{aligned}$$

The transmissibility is

$$\begin{aligned} T &= \left| \frac{1}{1 - \left(\frac{11.67\ \text{Hz}}{2.800\ \text{Hz}}\right)^2} \right| \\ &= 0.0611 \end{aligned}$$

The force transmitted to the structure is

$$\begin{aligned} F_t &= FT \\ &= (150\ \text{lbf})(0.0611) \\ &= 9.163\ \text{lbf} \quad (9.2\ \text{lbf}) \end{aligned}$$

The answer is (C).

Why Other Options Are Wrong

(A) This incorrect answer results from not converting the disturbing frequency from revolutions per minute to revolutions per second.

Depth

(B) This incorrect answer results from dividing by the deflection instead of its square root.

(D) This incorrect answer results from neglecting to square the frequency ratio in the transmissibility expression.

SOLUTION 61

According to the ASHRAE Fundamentals Handbook, the room criteria (RC) method has been preferred since at least 1981. This method has now been enhanced and is sometimes referred to as the revised RC Mark II method.

The answer is (A).

Why Other Options Are Wrong

(B) The A-weighted sound level method has limited usefulness, because it states acoustical performance as a single number.

(C) The noise criteria method has limited usefulness, because it states acoustical performance as a single number.

(D) The balanced noise criteria method has limited usefulness, because it states acoustical performance as a single number.

SOLUTION 62

The room sound pressure level at a chosen reference point is given by ASHRAE as

$$L_p = L_W - 5\log V - 3\log f - 10\log r + 25$$

The source sound power level (L_W) is 100 dB. The room volume is

$$\begin{aligned} V &= Lwh \\ &= (15\text{ ft})(15\text{ ft})(9\text{ ft}) \\ &= 2025\text{ ft}^3 \end{aligned}$$

The octave band center frequency, f, is 1000 Hz. The distance from the source to the observation point, r, is 12 ft.

$$\begin{aligned} L_p &= 100\text{ dB} - 5\log 2025 - 3\log 1000 - 10\log 12 + 25 \\ &= 88.7\text{ dB} \quad (89\text{ dB}) \end{aligned}$$

The answer is (C).

Why Other Options Are Wrong

(A) This answer incorrectly uses the natural rather than the common logarithm.

(B) This incorrect answer results from neglecting the constant 25.

(D) This incorrect answer results from neglecting the coefficients accompanying the logarithmic terms.

SOLUTION 63

The simple payback is the length of time it takes for the cumulative annual savings to equal the initial investment cost.

$$\text{payback period} = \frac{\text{initial investment}}{\text{net annual savings}}$$

In this case, the initial investment cost for the gas-fired heating system is not the full cost of the system, but the difference between the cost of the new system and the cost of repairing the existing system.

$$\begin{aligned} C_{\text{initial}} &= C_{\text{replace}} - C_{\text{repair}} \\ &= \$185{,}000 - \$40{,}000 \\ &= \$145{,}000 \end{aligned}$$

The existing electric resistance heating system can be considered to have a "site-based" heating efficiency of 100%. The annual heating cost of the existing system is

$$\begin{aligned} A_{\text{electric}} &= E_{\text{electric}} r_{\text{electric}} \\ &= \left(640{,}000\ \frac{\text{kW-hr}}{\text{yr}}\right)\left(\frac{\$0.08}{\text{kW-hr}}\right) \\ &= \$51{,}200/\text{yr} \end{aligned}$$

The natural gas-fired replacement system will increase the energy used based on its average annual fuel utilization efficiency (AFUE). The annual heating cost of the proposed system is

$$\begin{aligned} A_{\text{gas}} &= E_{\text{gas}}\left(\frac{1}{\text{AFUE}}\right) r_{\text{gas}} \\ &= \left(640{,}000\ \frac{\text{kW-hr}}{\text{yr}}\right)\left(\frac{1}{0.76}\right)\left(\frac{\$0.75}{\text{therm}}\right) \\ &\quad \times \left(3412\ \frac{\text{Btu}}{\text{kW-hr}}\right)\left(\frac{1\text{ therm}}{100{,}000\text{ Btu}}\right) \\ &= \$21{,}556/\text{yr} \end{aligned}$$

The annual savings for the gas system are

$$\begin{aligned} A_{\text{savings}} &= A_{\text{electric}} - A_{\text{gas}} \\ &= \frac{\$51{,}200}{\text{yr}} - \frac{\$21{,}556}{\text{yr}} \\ &= \$29{,}644/\text{yr} \end{aligned}$$

Thus,

$$\text{payback period} = \frac{\$145{,}000}{\frac{\$29{,}664}{\text{yr}}} = 4.89\ \text{yr} \quad (4.9\ \text{yr})$$

The answer is (B).

Why Other Options Are Wrong

(A) This answer results from incorrectly multiplying the annual energy usage for the gas system by the AFUE, rather than dividing by it.

(C) This answer results from incorrectly assuming the investment cost to be the total new system cost, rather than the difference between it and the cost of repairing the old system.

(D) This answer results from incorrectly dividing the investment cost by the alternative annual energy cost, rather than by the *difference* in annual cost between the fuels.

SOLUTION 64

The number of periods, n, needed to recover a present investment, P, with an annual recovery, A, at an interest rate, i, is

$$n = \frac{-\log\left(1 - \frac{Pi}{A}\right)}{\log(1+i)}$$

In this case, the recovery is the annual savings in energy costs made possible by the new boiler. This is the reduction in energy usage times the cost of energy.

$$A = (E_{\text{present}} - E_{\text{future}})r_{\text{gas}}$$

The future annual energy usage is the amount of energy consumed by the new boiler to produce the same amount of useful heat as the present boiler. This is the present energy usage multiplied by the ratio of present-to-future AFUE terms.

$$\begin{aligned} E_{\text{future}} &= \frac{E_{\text{present}}\text{AFUE}_{\text{present}}}{\text{AFUE}_{\text{future}}} \\ &= \frac{\left(200{,}000\ \frac{\text{therms}}{\text{yr}}\right)(0.75)}{0.84} \\ &= 178{,}571\ \text{therms/yr} \end{aligned}$$

The annual savings is

$$\begin{aligned} A &= \left(200{,}000\ \frac{\text{therms}}{\text{yr}} - 178{,}571\ \frac{\text{therms}}{\text{yr}}\right)\left(\frac{\$0.65}{\text{therm}}\right) \\ &= \$13{,}929/\text{yr} \end{aligned}$$

The number of years required to recover the boiler investment is

$$\begin{aligned} n &= \frac{-\log\left(1 - \frac{Pi}{A}\right)}{\log(1+i)} \\ &= \frac{-\log\left(1 - \frac{(\$85{,}000)(0.08)}{\frac{\$13{,}929}{\text{yr}}}\right)}{\log(1+0.08)} \\ &= 8.703\ \text{yr} \quad (8.7\ \text{yr}) \end{aligned}$$

The answer is (B).

Why Other Options Are Wrong

(A) This incorrect answer is the simple payback, without taking into account the time value of money.

(C) This incorrect answer results from neglecting to subtract the investment-to-savings ratio from 1 in the argument of the logarithm.

(D) This answer results from incorrectly multiplying present energy consumption by the difference in AFUE values, instead of by their ratio.

SOLUTION 65

The landlord wants a loan whose monthly payment is A for present value improvements of P. Therefore, the correct factor when multiplied by P must result in A. The left side of the equation must be A/P.

The needed formula is

$$\frac{A}{P} = \frac{i(1+i)^n}{(1+i)^n - 1}$$

The interest is compounded monthly, and payments will be monthly, so the number of periods (n) in 5 yr is 60. (Note that an annual interest rate of 6% is equivalent to a rate of 0.5% per month.)

The answer is (C).

Why Other Options Are Wrong

(A) This answer is incorrect because the left side should be A/P, and because there are 60 periods, not five.

(B) This answer is incorrect because there are 60 periods, not five. (It would be correct if the interest were compounded annually and the amount of an annual payment were sought.)

(D) This answer is incorrect because the left side should be A/P. (It would be correct if the landlord were starting with the monthly payment desired and wanted to calculate the amount of money that could be borrowed under these terms.)